LOCAL PROGRAMMING ON BROADCAST, CABLE, AND SATELLITE TELEVISION: STATUTES AND RULES

LOCAL PROGRAMMING ON BROADCAST, CABLE, AND SATELLITE TELEVISION: STATUTES AND RULES

CHARLES B. GOLDFARB

Nova Science Publishers, Inc.
New York

Copyright © 2008 by Nova Science Publishers, Inc.

All rights reserved. No part of this book may be reproduced, stored in a retrieval system or transmitted in any form or by any means: electronic, electrostatic, magnetic, tape, mechanical photocopying, recording or otherwise without the written permission of the Publisher.

For permission to use material from this book please contact us:
Telephone 631-231-7269; Fax 631-231-8175
Web Site: http://www.novapublishers.com

NOTICE TO THE READER

The Publisher has taken reasonable care in the preparation of this book, but makes no expressed or implied warranty of any kind and assumes no responsibility for any errors or omissions. No liability is assumed for incidental or consequential damages in connection with or arising out of information contained in this book. The Publisher shall not be liable for any special, consequential, or exemplary damages resulting, in whole or in part, from the readers' use of, or reliance upon, this material.

Independent verification should be sought for any data, advice or recommendations contained in this book. In addition, no responsibility is assumed by the publisher for any injury and/or damage to persons or property arising from any methods, products, instructions, ideas or otherwise contained in this publication.

This publication is designed to provide accurate and authoritative information with regard to the subject matter covered herein. It is sold with the clear understanding that the Publisher is not engaged in rendering legal or any other professional services. If legal or any other expert assistance is required, the services of a competent person should be sought. FROM A DECLARATION OF PARTICIPANTS JOINTLY ADOPTED BY A COMMITTEE OF THE AMERICAN BAR ASSOCIATION AND A COMMITTEE OF PUBLISHERS.

LIBRARY OF CONGRESS CATALOGING-IN-PUBLICATION DATA
Goldfarb, Charles B.
  Local programming on broadcast, cable, and satellite television : statutes and rules / Goldfarb, Charlies B.
    p. cm.
  ISBN 978-1-60456-276-7 (softcover)
  1. Television--Law and legislation--United States. 2. Cable television--Law and legislation--United States. 3. Direct broadcast satellite television--Law and legislation--United States. I. Goldfarb, Charles B. II. Title.
KF2840.A3 2007
343.7309'946--dc22
                                                              2007048144

Published by Nova Science Publishers, Inc.
New York

CONTENTS

Preface		vii
Chapter 1	Introduction	1
Chapter 2	Broadcast Television	5
Chapter 3	Cable Television	11
Chapter 4	Satellite Television	21
Chapter 5	Issues for Congress	27
References		77
Index		89

PREFACE

Most broadcast television stations' viewing areas extend far beyond the borders of their city of license, and in many cases extend beyond state borders. Under existing FCC rules, which are intended to foster "localism," the licensee's explicit public interest obligation is limited to serving the needs and interests of viewers within the city of license. Yet, in many cases, the population residing in the city of license is only a small proportion of the total population receiving the station's signal. Hundreds of thousands of television households in New Jersey (outside New York City and Philadelphia), Delaware (outside Philadelphia), western Connecticut (outside New York City), New Hampshire (outside Boston), Kansas (outside Kansas City, Missouri), Indiana (outside Chicago), Illinois (outside St. Louis), and Kentucky (outside Cincinnati) have little or no access to broadcast television stations with city of license in their own state. The same holds true for several rural states including Idaho, Arkansas, and especially Wyoming, where 54.55% of television households are located in television markets outside the state. Although market forces often provide broadcasters the incentive to be responsive to their entire serving area, that is not always the case. This report provides, for each state, detailed county-by-county data on the percentage of television households located in television markets outside the state and whether there are any in-state stations serving those households. The Nielsen Designated Market Areas ("DMAs") also often extend beyond state borders. Local cable operators are required to carry the broadcast signals of television stations located in their DMA. If they are located in a DMA for which the primary city is in another state, and most or all of the television stations in that DMA have city of license in the other state, then the broadcast television signals they must carry will be primarily or entirely from out of state. In some cases, they may not be allowed to carry signals

[1]from within the state but outside the DMA to provide news or sports programming of special interest in their state because of network nonduplication, syndicated exclusivity, or sports programming blackout rules or because of private network affiliation contract agreements, or may be discouraged to do so because these signals do not qualify for the royalty-free permanent compulsory copyright license for local broadcast signals. The Satellite Home Viewer Extension and Reauthorization Act of 2004 expanded the scope of in-state television signals that satellite operators are permitted (and in some cases required) to offer subscribers. In addition to the signals of those broadcast television stations with city of license within the DMA in which the subscriber is located ("local-into-local" service), satellite operators may offer (subject to certain restrictions) signals from outside the DMA if those signals are "significantly viewed" by those households in the subscriber's geographic area that only receive their broadcast signals over-the-air (not via cable or satellite). In addition, satellite operators may offer certain subscribers located in New Hampshire, Vermont, Mississippi, and Oregon certain in-state signals from outside the subscribers' DMA and must offer subscribers in Alaska and Hawaii certain in-state signals.

---

[*] Excerpted from CRS Report RL32641, dated July 17, 2007.

Chapter 1

INTRODUCTION

Many Members of Congress receive complaints from constituents that the news, information, and even entertainment television programming available to them does not address the needs and interests of their local community. These constituents question, for example, why they cannot receive local news programming that focuses on the issues of importance in their locality or state or the football games of their state university.

Sometimes desired programming cannot be provided because of private contractual network affiliation agreements between broadcast networks and local broadcast station affiliates. But at other times desired programming cannot be provided because the geographic boundaries of broadcast signal contours, audience viewing patterns, and governmental jurisdictions do not conform with one another; as a result, no methodology for allocating broadcast spectrum or for constructing rules about which viewers a broadcaster's programming must serve or which signals cable and satellite operators must or may carry will meet the needs of all viewers or communities.

For example, millions of U.S. television households are located in the same metropolitan area as a major city, but across state lines from that city. Some of those households will have a stronger affinity for programming that focuses on issues relevant to the major city; others will have a stronger affinity for programming that focuses on relevant state issues. Using either metropolitan area hubs or state borders as the basis for determining the programming obligations of stations whose signals reach beyond city borders and state borders will inevitably disappoint some households. With or without government intervention, it is inevitable that some viewers and some communities will feel their needs and interests are not being met. At the same time, it may be possible to make the existing statutes and rules that affect the television programming available to consumers more flexible in order to foster the provision of television programming that better meets the needs of local communities. The purpose of this report is to explain how existing statutes and rules affect the television programming available to consumers and to discuss potential ways to foster the provision of television programming that better meets the needs of local communities.

Each broadcast television license is assigned a community of license, in the form of a specific city. Most broadcast television stations' viewing areas extend far beyond the borders of their city of license, and in many cases extend beyond state borders.

The local broadcast television stations that each cable system must carry are determined by the Nielsen Designated Market Area (DMA) in which the cable system is located. In the 1992 Cable Act, Congress amended the 1934 Communications Act to require, subject to certain exceptions, each cable system to carry the signals of all the local full power commercial television stations "within the same television market as the cable system," with that market determined by "commercial publications which delineate television markets based on viewing patterns."[1]

The DMAs represent the only nationwide commercial mapping of television audience viewing patterns. Each county in the United States is assigned to a television market based on the viewing habits of the residents in the county.[2] Since viewing patterns are more closely aligned with the economic markets in which households participate than with state boundaries, some counties are assigned to DMAs for which the primary city is in a different state. In a DMA that straddles two states, with the major city and most of the broadcast stations located in one state, the cable systems in the other state may find that few or none of the broadcast station signals they must carry are from their own state.

Until Congress passed the Satellite Home Viewer Extension and Reauthorization Act (SHVERA)[3] in November 2004, a satellite system, when providing local service, could offer a subscriber only the signals of those local broadcast stations located within the same DMA as the subscriber (called "local-into-local" service); it was prohibited from offering broadcast signals that might have originated nearby but outside the subscriber's DMA.[4] As a result, in many situations, those subscribers to satellite service who were located in DMAs in which all the broadcast television stations are in another state (typically because the primary city in the DMA is in another state) could not be provided the signals of any in-state local broadcast television stations.[5] SHVERA expanded the scope of in-state television signals that satellite operators are permitted (and in some cases required) to offer subscribers. In addition to the signals of those broadcast television stations with city of license within the DMA in which the subscriber is located, satellite operators may offer (subject to certain limitations) signals from outside the DMA if those signals are "significantly viewed" by those households in the subscriber's geographic area that only receive their broadcast signals over-the-air (not via cable or satellite).[6] In addition, under SHVERA, satellite operators may offer certain subscribers located in New Hampshire, Vermont, Mississippi, and Oregon certain in-state signals from outside the subscribers' DMA and must offer subscribers in Alaska and Hawaii certain in-state signals.[7]

Table 1, which is appended to this report, presents a compilation of data from Nielsen Media Research[8] and *Television & Cable Factbook 2004[9]* on the number and location of U.S. television households that are located in DMAs for which the primary city is in a different state. It identifies, for each state:

- the number of television households in the state;
- the counties in the state assigned to DMAs for which the primary city is outside the state;
- the number of television households in those counties;
- the percentage of television households in the state that are located in DMAs for which the primary city is outside the state; and
- the full power broadcast television stations with city of license or transmitting location inside the state that are located in DMAs for which the primary city is outside the state.

These data provide the empirical basis for the discussion in this chapter. The information on city of license in table 1 is very important. It shows whether television households in counties assigned to DMAs for which the primary city is outside the state nonetheless have in-state, in-DMA television stations available to them.[10] For example, it shows that the approximately 55,000 Arkansas television households that are located in the Springfield, Missouri DMA receive service from two UHF analog[11] stations in their DMA that have city of license in Arkansas and that therefore have the obligation to meet the needs and interests of Arkansas viewers. But approximately 77,000 Arkansas households that are located in the Memphis, Tennessee DMA have access to no broadcast television stations that have city of license in Arkansas (and thus have no access to broadcast television stations that have an obligation to serve the needs and interests of those

Arkansas households). What table 1 does not provide is information about whether and how well the needs and interests of these television households are being met by broadcast television stations with city of license in the other state (for example, how well the broadcast stations with city of license in Memphis, Tennessee are serving the needs and interests of those 77,000 television households in Arkansas).

Chapter 2

BROADCAST TELEVISION

Localism has long been one of the three primary objectives of U.S. broadcast policy.[12] Broadcasters are considered to be temporary trustees of the public's spectrum because the 1934 Communications Act instructs the Federal Communications Commission (FCC or Commission) to award licenses to use the airwaves expressly on the condition that licensees serve the public interest;[13] section 309(a) requires the Commission to determine, in the case of applications for licenses, "whether the public interest, convenience, and necessity will be served by granting such application."[14] As trustees of the public airwaves, broadcasters must serve the public interest by airing programming that is responsive to the interests and needs of their community of license. The concept of localism derives from Title III of the Communications Act; section 307(b) of the act explicitly requires the Commission to "make such distribution of licenses, frequencies, hours of operation, and of power among the several States and communities as to provide a fair, efficient, and equitable distribution of radio service to each of the same."[15]

In carrying out the mandate of Section 307(b), when the Commission allocates channels for a new broadcast service, its first priority is to provide general service to an area, but its next priority is for facilities to provide the first local service to a community.[16] The Commission has long recognized that "every community of appreciable size has a presumptive need for its own transmission service."[17] The Supreme Court has stated that "[f]airness to communities [in distributing radio service] is furthered by a recognition of local needs for a community radio mouthpiece."[18]

Once awarded a license, a broadcast station must place a specified signal contour over its community of license to ensure that local residents receive service.[19] A station must maintain its main studio in or near its community of license to facilitate interaction between the station and the members of the local community it is licensed to serve.[20] In addition, a station "must equip the main studio with production and transmission facilities that meet the applicable standards, maintain continuous program transmission capability, and maintain a meaningful management and staff presence."[21] The main studio also must house a public inspection file, the contents of which must include "a list of programs that provided the station's most significant treatment of community issues during the preceding three month period."[22]

In practice, full power broadcast television signal contours almost always extend far beyond the borders of the community (city) of license. For a full power television station, the geographic boundaries of its city of license are narrower than the geographic area that can receive the signals of the station. As shown in figure 1, the Grade B contour for a hypothetical full power broadcast television station licensed to serve major city M, in state X, extends far beyond the borders of that city, and even into state Y.[23]

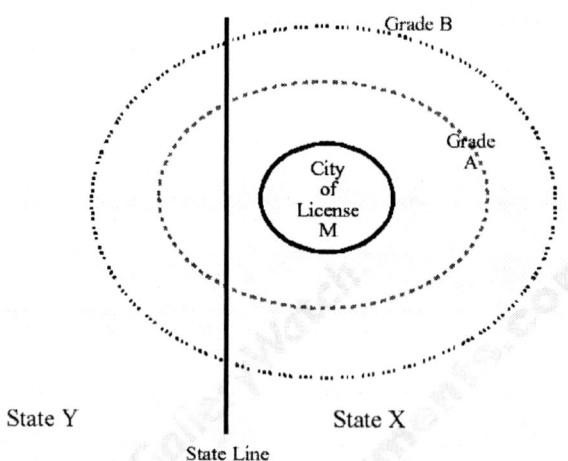

Source: CRS.

Figure 1. Broadcast television station city of license and signal reach.

Under existing FCC rules, the licensee's explicit public interest obligation is limited to serving the needs of viewers within its city of license.[24] Over the years, the Commission has interpreted its rules to carry a secondary obligation for the licensee to serve the needs of viewers outside the city of license but within the signal reach.[25]

But the FCC rules do not provide specific guidance about this secondary obligation. Yet, in many cases, the population residing within the city of license is only a small proportion of the total population receiving the station's signal.

Many broadcast television stations have viewing areas that cross state borders. This is not surprising as cities often are located along rivers or other natural boundaries that act as state borders, but urban development often occurs on both sides of the border and a station's viewing radius around a central city will extend into suburbs and even into other cities across state borders. In many of these situations, the FCC has attempted to serve populations on both sides of the state borders by assigning some licenses to cities in each state. For example, in the Paducah, Kentucky-Cape Girardeau, Missouri-Mount Vernon, Illinois area, an NBC-affiliated VHF analog and digital station and a UHF analog and digital station have city of license in Paducah, Kentucky, a Fox-affiliated UHF analog and digital station and the CBS-affiliated VHF analog and digital station have city of license in Cape Girardeau, Missouri, an ABC-affiliated VHF analog and digital station has city of license in Harrisburg, Illinois, and a VHF analog station has city of license in Mount Vernon, Illinois.

But some metropolitan areas are dominated by a single large city, with most of the television licenses (including those for all the stations affiliated with the four major broadcast networks) assigned to that city and the licenses for only a few stations assigned to cities in the neighboring state(s). For example, the licenses for the preponderance of stations serving the metropolitan New York City and Philadelphia areas are assigned to those cities, with very few licenses assigned to New Jersey, western Connecticut, or Delaware. The FCC has taken special notice of this situation with respect to the state of New Jersey by explicitly stating that all the New York and Philadelphia stations have the responsibility to serve the needs of their New Jersey viewers.[26]

As a result of television viewing areas extending beyond state borders, and the frequency of populations being concentrated along both sides of those borders, there are a number of situations in which, despite the efforts of the FCC, a significant proportion of the television households in a state are served primarily or entirely by broadcast television stations whose city of license — and hence primary service obligation — lies outside that state.[27]

There are 3,149,060 television households in New Jersey, but the vast majority of these households receive all or most of their over-the-air signals from stations licensed to New York City or Philadelphia. Only one television station with a city of license in New Jersey is affiliated with one of the major television networks, and that UHF analog NBC-affiliate serves the relatively sparsely populated southern tip of the state. Of the other eight commercial stations with city of license in New Jersey, only one is a VHF station and five are affiliated with Spanish language networks. Four of the New Jersey stations transmit from locations in New York City and their signals fully cover that city.[28] As indicated earlier, at least at the policy level, the Commission has attempted to address the potential lack of coverage of New Jersey-specific issues by explicitly requiring the stations licensed to New York City and Philadelphia to offer programming that serves the New Jersey households within their viewing areas.

There are 313,630 television households in Delaware, but there is only one UHF analog commercial station with city of license in the state, plus three UHF noncommercial stations. The bulk of the Delaware population is served by television stations in Philadelphia; those stations have primary obligations to serve the viewers of Philadelphia and suburban New Jersey. The remainder of the Delaware viewers are served by stations in Salisbury, Maryland. Similarly, the only major network with an affiliate in New Hampshire is ABC. The vast majority of New Hampshire's 498,150 television households receive broadcast service primarily from stations in Boston.

This pattern exists around many large cities. More than 900,000 television households in Maryland are in the Washington, DC DMA. Although a small portion of these households are served by a UHF analog and digital ABC-affiliated station and a UHF analog independent station, both with city of license in Hagerstown, Maryland, and a UHF analog noncommercial station with city of license in Frederick, Maryland, most are primarily served by Washington, DC stations. The Baltimore stations provide a potential source of programming that addresses Maryland-specific issues, but although most of these households fall within those stations' Grade B contours, most households subscribe to cable or satellite service and therefore few of them have the antennas needed to bring in the Baltimore stations.[29] While Washington, DC stations do address issues of interest to Maryland suburbanites, they have the burden of addressing the needs of three jurisdictions, with primary obligations to serve DC. In Virginia, as well, more than 900,000 television households are in the Washington, DC DMA, and are served primarily by Washington, DC stations, with only a UHF analog and digital independent station, a UHF analog Telefuturaaffiliated station, a UHF analog and digital noncommercial station, and two UHF analog noncommercial stations located in that portion of Virginia.

More than 300,000 Kansas television households are in the Kansas City, Missouri DMA and rely almost entirely on broadcast stations from that city. The only station in that DMA with city of license in Kansas is a UHF analog and digital station in Lawrence, Kansas. The Kansas City stations do not have explicit obligations to meet the needs of their Kansas viewers. Similarly, more than 330,000 — or just under 35% — of the television households in Connecticut are in the New York City DMA and primarily served by New York City stations; more than 150,000 Kentucky television households are in the Cincinnati, Ohio DMA and there are no commercial stations in that DMA with city of license in Kentucky; almost 200,000 television households in northwestern Indiana are in the Chicago, Illinois DMA, served primarily by Chicago stations, with only one UHF analog and digital commercial station and one UHF analog noncommercial station located in that part of Indiana; and more than 300,000 television households in western Illinois are in the St. Louis, Missouri DMA, served primarily by St. Louis stations, with only one UHF analog and digital commercial station with city of license in Illinois.

This problem is not limited to major metropolitan areas. As shown in table 1, 54.55% of the television households in Wyoming are located in television markets outside the state. The population centers around Casper and Cheyenne are served by broadcast stations with city of license in Wyoming, but most other parts of the state are served primarily or entirely by broadcast stations with city of license outside the state. Almost one-fourth of Idaho's television households are in DMAs whose principal city is outside the state and more than one-fifth of Arkansas' television households are in DMAs whose principal city is outside the state.

Chapter 3

CABLE TELEVISION

As early as the 1960s, when households began receiving their broadcast signals over cable television rather than over the air, Congress became concerned both that local broadcasters could be harmed (either because they were not compensated for their programming or because local cable systems chose not to carry their programming, thus cutting off their access to a large segment of the viewing audience) and that there could be a diminution of programming that serves local needs and interests. Congress therefore enacted several laws intended to extend the policy goal of localism to the cable television industry, including the 1972, 1984, and 1992 Cable Acts.

'MUST CARRY' RULES

Most notable was the adoption of the "retransmission consent/must carry" election in the 1992 Cable Act. Every three years, each local commercial broadcast television station must choose between:

- negotiating retransmission consent agreements with the cable systems operating in its service area, whereby if agreement is reached the broadcaster is compensated by the cable system for the right to carry the broadcast signal, and if agreement is not reached, the cable system is not allowed to carry the signal; or
- requiring each cable system operating in its service area to carry its signal, but receiving no compensation for such carriage.

With this mandatory election, broadcasters with popular programming that are confident the local cable systems will want to carry that programming can make the retransmission consent election and be assured compensation for such carriage, and broadcasters with less popular programming that the local cable systems might otherwise not choose to carry can make the must carry election and be assured that their signal will be carried by all local cable systems.

The evolution of the must carry rules demonstrates the difficulty of constructing rules that safeguard local broadcasters and foster local programming without unduly burdening cable operators or undermining the exclusive distribution contracts between program content providers and program distributors. The initial rules required cable operators to carry all broadcast television signals whose Grade B signals reached into the cable service area. But this proved too expansive; for example, the Grade B contours of Washington, DC stations extend over Baltimore, and vice versa. The must carry requirements were then scaled back to those signals from stations located within certain mileage limits (for example, within 35 miles). There then was some concern that this would harm broadcast stations that did not meet these mileage limits but had historically been viewed by audiences beyond those mileage limits. The must carry rules were modified to apply to all broadcast stations that were "significantly viewed" by those households in the cable service area that did not receive service from cable or satellite providers.[30] The specific threshold viewing levels were, for a network-affiliate station, a market share of at least 3% of total weekly viewing hours in the market and a net weekly circulation of 25%; for independent stations, 2% of total weekly viewing hours and a net weekly circulation of 5%. The share of viewing hours referred to the total hours that households that do not receive television signals from multichannel video program distributors ("MVPDs")[31] viewed the subject station during the week, expressed as a percentage of the total hours these households viewed all stations during the week. Net weekly circulation referred to the number of households that do not receive television signals from multichannel video programming distributors that viewed the station for five minutes or more during the entire week, expressed as a percentage of the total households that do not receive television signals from multichannel video programming distributors in the survey area.

But as more and more households subscribed to cable service, it became less reasonable to base must carry decisions on the behavior of the minority of households that continued to get their service over the air. In the 1992 Cable Act, Congress modified sections 614 and 615 of the 1934

Communications Act[32] to base the must carry rules on a definition of local television markets explicitly based on viewing patterns, requiring each cable operator to carry the signals of local commercial television stations, qualified low power stations, and qualified noncommercial educational stations, if the licensees of those stations chose to have their signals carried.[33] This statutory language remains in place today.

The exact number of broadcast signals that cable systems must carry varies with the size of the cable system, but includes at a minimum three local commercial stations and one local noncommercial educational station. Cable systems with more than 12 channels must carry local commercial broadcast stations on up to one-third of their channels and up to three qualified noncommercial educational stations.

"Local" commercial stations are defined as all stations whose community of license is within the same television market as the cable system.[34] Following the statutory directive to use television markets delineated by commercial publications, the FCC implemented a rule defining television markets according to the Nielsen DMAs.[35]

OTHER FEDERAL RULES AND LAWS

Cable systems must carry the entirety of the program schedule of every local television station carried pursuant to the mandatory carriage provisions (or the retransmission consent provisions) of the 1992 Cable Act, subject to the carriage restrictions in the network program non-duplication rules,[36] syndicated exclusivity protection rules,[37] and sports programming blackout rules.[38] In practice, these rules are quite complex and result in significant amounts of programming from television stations within a cable operator's DMA not being carried because such carriage would be duplicative or would contravene exclusivity agreements.

Interestingly, while the must carry rules are now based on DMAs, the non-duplication rules continue to be based on the old "significantly viewed" criteria. Consider a cable operator that sought to carry the broadcast signals of a network-affiliated station that is located nearby, but outside the DMA in which the cable system is located, and that successfully worked out a retransmission consent agreement with that affiliated station. For example, assume a Montgomery County, Maryland, cable operator, which is located in the Washington, DC DMA, sought to carry a Baltimore broadcast station, and successfully worked out a retransmission consent agreement with the Baltimore station. Then, if that Baltimore station met the "significantly viewed" criteria in the cable operator's location, its signals would not be subject to the non-duplication rules and the signals from both the Washington, DC network affiliate and the Baltimore network affiliate could be carried by the Montgomery County cable operator in their entirety, without blackouts of the network programming on the Baltimore station. Some industry observers claim, however, that such duplication does not occur very often because the national networks, rather than the affiliated stations, tend to make the determination (through language in the private contractual agreement between the network and each affiliate) about whether a station located outside a cable system's DMA should grant the cable system retransmission consent — and frequently these contracts effectively preclude retransmission consent.

Copyright law also may tend to discourage cable systems from carrying the signals of broadcast stations located outside the DMA in which the cable system is located.[39] Cable systems are required to pay royalties under a congressionally granted compulsory copyright license for the "secondary transmission" of the signals of broadcasters located outside the DMA within which the cable system is located. In contrast, cable systems enjoy a royalty-free permanent compulsory copyright license — that is, do not have to pay copyright fees — for the secondary transmission of broadcast signals of stations located in their DMAs. The royalty-free license extends to the secondary transmission of signals of out-of-DMA broadcast stations that meet the "significantly viewed" criteria discussed above. However, if an in-state, but out-of-DMA station does not meet the "significantly viewed" criteria, the requirement to pay the copyright royalties might tip the balance away from the cable system carrying the station's signals.

FLEXIBILITY IN THE RULES

The 1992 Cable Act includes explicit language authorizing the FCC to implement the must carry rules flexibly in order to foster the goal of localism. The language in Section 614(h)(1)(C) of the act (Carriage of Local Commercial Signals)[40] explicitly allows for exceptions, requiring the carriage of "local commercial television stations," but providing flexibility on how those local stations would be determined:

For purposes of this section, a broadcasting station's market shall be determined by the Commission by regulation or order using, where available, commercial publications which delineate television markets based on viewing patterns, except that, following a written request, the Commission may, with respect to a particular television broadcast station, include additional communities within its television market or exclude communities from such station's television market to better effectuate the purpose of this section. In considering such requests, the Commission may determine that particular communities are part of more than one television market.

In considering requests filed pursuant to clause (i), the Commission shall afford particular attention to the value of localism by taking into account such factors as —

- whether the station, or other stations located in the same area, have been historically carried on the cable system or systems within such community;
- whether the television station provides coverage or other local service to such community;
- whether any other television station that is eligible to be carried by a cable system in such community in fulfillment of the requirements of this section provides news coverage of issues of concern to such community or provides carriage or coverage of sporting and other events of interest to the community; and
- evidence of viewing patterns in cable and noncable households within the areas served by the cable system or systems in such community.

In a 2001 decision involving the attorney general of the state of Connecticut, the Commission found that only a broadcaster or a cable system has the standing to file a request to modify the signal carriage right of a broadcast station.[41]

THE DIGITAL TRANSITION AND LOCAL PROGRAMMING

The television industry is in the midst of another policy debate involving cable carriage of local broadcast signals during (and after) the congressionally mandated transition from analog transmission of broadcast signals to digital transmission.[42]

During the transition, television broadcasters have been given additional spectrum to allow them to broadcast using digital technology while retaining the spectrum they use for analog broadcasting. The Deficit Reduction Act of 2005 (P.L. 109-171) set the digital transition deadline at February 17, 2009, by which date the broadcasters will be required to return the spectrum used for analog transmission. During this transition, many broadcasters are providing both analog and digital broadcast signals. Therefore there has been a public policy debate over which broadcast signals cable systems should be obligated to carry. In January 2001, the FCC announced adoption of rules for cable carriage of digital television signals. The FCC ruling does not require cable systems to simultaneously carry both the analog and digital signals ("dual carriage") of local television stations. The FCC tentatively concluded that "such a requirement appears to burden cable operators' First Amendment interests more than is necessary to further a substantial governmental interest." While not approving a dual carriage mandate, the FCC did rule that a digital-only television station, whether commercial or non-commercial, can immediately assert its right to carriage on a local cable system. In addition, a television station that returns its analog spectrum and converts to digital operations must be carried by local cable systems.

In April 2007, the FCC issued a notice of proposed rulemaking to address another issue: how to protect those households that subscribe to cable systems that have not fully deployed digital technology by the February 17, 2009, deadline for broadcasters to discontinue analog transmission.[43] The Commission addressed the statutory requirement that cable operators must make the signal transmitted by a broadcaster electing mandatory carriage viewable by all of their subscribers,[44] seeking comment on how cable operators can implement this requirement after the end of analog broadcasting. The Commission proposed that cable operators must comply with this "viewability" provision and ensure that cable subscribers with analog television sets are able to continue to view all must-carry stations after the end of the digital television transition by either (1) carrying the digital signal in analog format, or (2) carrying the signal only in digital format, provided that all subscribers have the necessary equipment to view the broadcast content. Although all the commissioners agreed that this was an important issue that the Commission should address, two of the commissioners raised several questions: whether it was premature to propose specific prescriptive rules in light of potentially strong market forces that could resolve any problem, whether all the constitutional issues had been fully vetted, whether it was premature to propose reversal of existing decisions without having had a chance to get public comment.[45]

Cable systems must carry "primary video," defined as a "single programming stream and other program-related content." With digital technology, broadcasters can divide their 6 MHz of spectrum into separate and discrete streams of content and broadcast multiple (as many as six) channels of programming. This is known as "multicasting." Broadcasters sought an FCC ruling requiring cable operators to carry any and all multicasted channels transmitted by commercial broadcasters, arguing that the incentive to develop additional programming streams is diminished if they have no guarantee that cable systems will carry that programming. Cable providers countered that their decision on whether or not to carry additional broadcaster programming streams should be dictated by the market, not mandated. In February 2005, the FCC affirmed its prior decision that cable operators are not required to carry more than a single digital programming stream from any particular broadcaster.[46]

LOCAL FRANCHISE REQUIREMENTS

Under the 1984 Cable Act, local franchising authorities may require cable operators to set aside channels for public, educational, or governmental (PEG) use.[47] In addition, franchising authorities may require cable operators to provide services, facilities, and equipment for the use of these channels. Many cable systems include several PEG channels. In general, cable operators are not permitted to control the content of programming on PEG channels. Cable operators may impose non-content-based requirements, such as minimum production standards, and may mandate equipment user training. In addition, cable systems may make available "access channels" that typically provide community-oriented programming, such as local news, public announcements and government meetings. They are usually programmed by individuals or groups, on either public, educational or governmental access channels or on commercial leased access channels.

SUMMARY OF FACTORS AFFECTING LOCAL PROGRAMMING ON CABLE

The bottom line of the existing rules is as follows. Unless they have systems with very small capacity, local cable operators are required to carry the broadcast signals of all the full-power television stations (and certain qualified low power television stations) located in their DMA and non-commercial stations that transmit from within 50 miles of the cable head-end whose grade B contours cover the cable system's service area. The cable operator is required to carry the entirety of the program schedule of each of these broadcast stations (subject to possible blackouts to conform with the non-duplication rules for those circumstances where more the broadcast programming is duplicated by a second or third station in the DMA).

If a cable system is located in a DMA in which the primary city is in another state, and most or all of the television stations in that DMA have city of license in the other state, then the broadcast television signals it must carry will be primarily or entirely from out of state. This scenario is shown in figure 2. Although local cable operator Q's franchise is located in state Y, and the major nearby city, M, is located in state X, both are within the same DMA, F. If local cable operator Q wants to carry the signals of broadcasters that are located in state Y but outside of DMA F, it can negotiate with those broadcasters to carry their signals, but any carriage would be subject to the restrictions in the network program non-duplication, syndicated programming exclusivity protection, and sports programming blackout rules, and to copyright fees (though these rules and fees will not be in effect if the "significantly viewed" criteria can be met). All these factors may restrict the state-specific entertainment programming cable operator Q can carry and also could affect the local news programming carried. Cable operator Q is not likely to use one of its channels to offer a "Swiss cheese" program schedule with holes in it for blacked out programs or programs for which it does not choose to pay copyright fees. Nor is it likely to set aside a channel just for several hours a day of state news or one or two sports events per week.

Some observers claim, however, that when cable operators do not carry the instate programming of out-of-DMA broadcast signals, it is unlikely to be because of these rules, which frequently can be sidestepped through application of the exceptions for "significantly viewed" stations. Rather, these observers claim, it is likely to be because the in-state broadcasters are constrained by territorial exclusivity provisions in their network affiliation agreements, allegedly imposed by the broadcast networks.

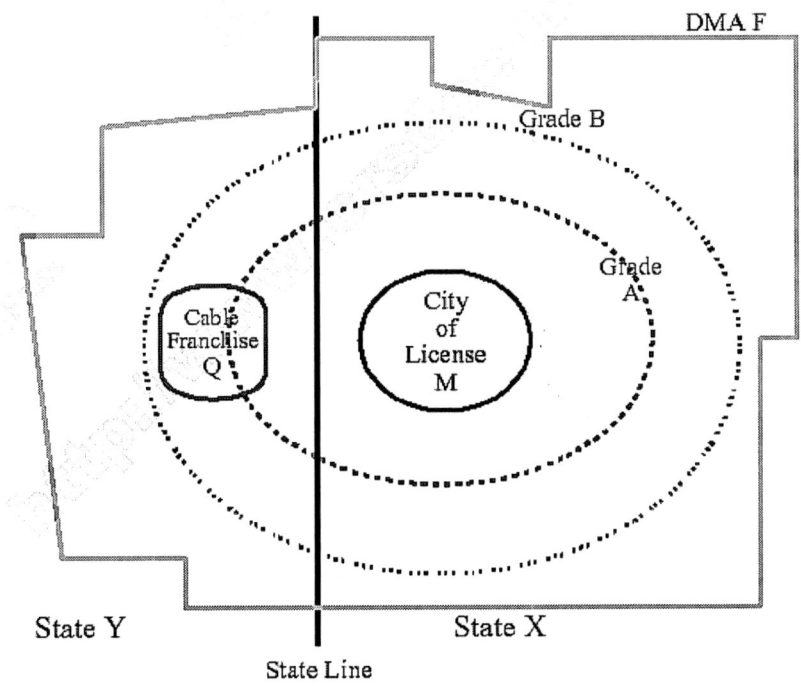

Source: CRS.

Figure 2. A Cable System. Located in a DMA in which the primary city is in another state.

Whatever the cause of cable system reluctance to carry the signals of in-state, but outside-the-DMA broadcast signals, it is likely that the within-DMA, but out-of-state broadcasters (for example, the broadcasters with city of license M) will cover some issues of interest to the cable operator's subscribers (the subscribers to cable system Q). Their inclusion in the same DMA is based on the assumption that viewers in the county in which the cable system operates tend to view the signals from that DMA and are likely to have a marketplace connection that broadcasters will have an incentive to foster. But the coverage of issues specific to the viewers in that cable system's service area may be quite limited since the broadcasters are not subject to any explicit obligation to serve the needs of viewers outside their city of license and their close-in viewers are likely to be considered more valuable by advertisers.

Whether or not this represents a problem to the cable system's subscribers will depend on their relative affinity toward news, information, and sports programming focused on the television market in which they are located, as defined by the DMA, vs. news, information, and sports programming focused on the political jurisdiction (state) in which they reside. For example, a cable subscriber in Montgomery County, Maryland, might have a preference for programming from Washington, DC stations that presents detailed traffic information on commuter routes between the subscriber's home and downtown Washington or, alternatively, might have a preference for programming from Baltimore, Maryland stations that presents more in-depth reporting of Maryland state politics. The current rules assume the preference is for the former because it is based on the statutory requirement that must carry requirements mirror existing viewing patterns.

Chapter 4

SATELLITE TELEVISION

Until Congress passed the Satellite Home Viewer Improvement Act (SHVIA) of 1999,[48] satellite television providers were not allowed to provide local broadcast television signals to their subscribers. SHVIA sought to promote competition between cable television and direct broadcast satellite, and to increase local program choices available to television households, by allowing satellite companies to provide local broadcast television signals to all subscribers who reside in the local television station's market. Local markets are explicitly defined in the statute as the Nielsen DMAs. This ability of satellite companies to provide local broadcast channels is commonly referred to as "local-into-local" service. Satellite companies are not required to offer local-into-local service, and they can charge for the service. Under copyright law, satellite companies enjoy a royalty-free permanent compulsory copyright license — exempting them from paying copyright royalties — for the secondary transmission of the broadcast signals of stations provided to subscribers as part of local-into-local service (the signals of broadcast stations located in the DMA of the subscriber).[49] But if a satellite system chooses to provide local-into-local service in any DMA, it must provide subscribers in that DMA with all of the local broadcast television signals that are assigned to the DMA that ask to be carried on that satellite system. A satellite system is not required to carry more than one local broadcast television station that is affiliated with a particular television network unless the stations are licensed to communities in different states.[50]

Under SHVIA, local-into-local service was explicitly restricted by law to the provision of the signals of broadcast television stations with city of license within the DMA in which the customer is located. Satellite operators did not have the opportunity that cable operators have to negotiate carriage of the programming of broadcasters that are in-state, but outside the viewer's DMA, unless the satellite operator's customers were unable to receive over-the-air broadcast signals of a Grade B intensity and therefore qualified, under a different section of law,[51] to receive distant network signals that may be (but need not be) from within state. This situation is shown in figure 3. Satellite subscriber Z is located in state Y and in DMA F. Under SHVIA, the satellite operator could provide subscriber Z local-into-local service consisting only of the signals of broadcast television stations located in DMA F, even if none of those stations are located in state Y. Nor could the satellite operator offer subscriber Z any distant network signals that originated from state Y because subscriber Z is within the Grade B contour of the broadcast stations in city M. Because of these rules, news or sports entertainment that was broadcast by a station in central Wyoming or Arkansas often was not available to satellite subscribers in more remote parts of those states that were within out-of-state DMAs.

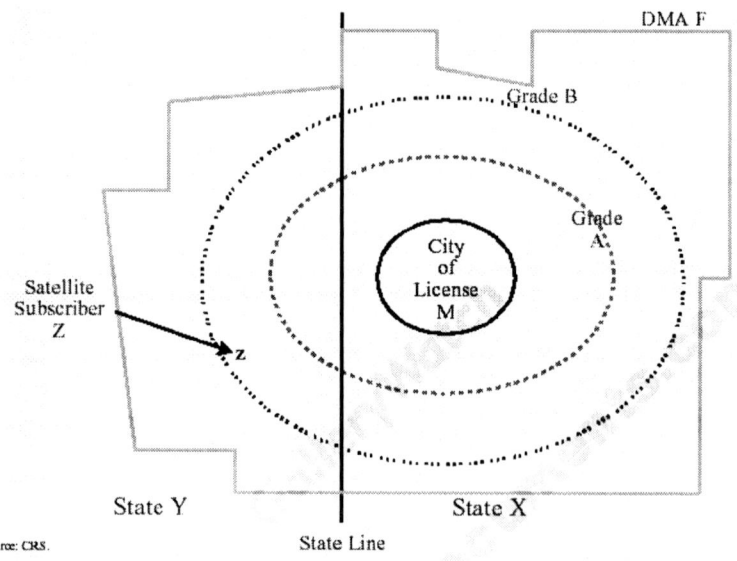

Figure 3. Satellite subscriber whose local broadcast television stations, as defined by the DMA, are in a different state.

This restriction on local-into-local service was not based on technological constraints or lack of bandwidth (although the number of DMAs in which local-into-local service is offered may be affected by bandwidth and satellite capacity constraints). Once a satellite operator has uplinked the programming of a particular broadcast station to a satellite, there are no technical constraints on making that signal available to all television households within the footprint of the satellite. (It is true, however, that the greater use of spot beams has resulted in smaller footprints so there may now be situations in which the broadcast signal of a station in a particular state is uplinked to a satellite with a spot beam that does not cover other portions of the state that are located in a different DMA.) But in most cases, the primary reason why a subscriber did not receive broadcast signals from stations located outside that subscriber's DMA was that the satellite operator, in order to conform with the law, had to set the subscriber's set-top box to exclude the out-of-DMA signals emanating from its satellite.

The Satellite Home Viewer Extension and Reauthorization Act of 2004 (SHVERA) expanded the scope of in-state television signals that may be of local interest to subscribers that satellite operators are permitted (and, in the case of operators in Alaska and Hawaii, required) to offer subscribers. In addition to the signals of those broadcast television stations with city of license within the DMA in which the subscriber is located ("local-into-local" service), satellite operators may offer (subject to certain limitations) signals from outside the DMA if those signals are "significantly viewed" by those households in the subscriber's geographic area that only receive their broadcast signals over-the-air (not via cable or satellite). In addition, satellite operators may offer certain subscribers located in New Hampshire, Vermont, Mississippi, and Oregon certain in-state signals from outside the subscribers' DMA and must offer subscribers in Alaska and Hawaii certain in-state signals.

Specifically, the current restrictions on the retransmission of distant broadcast signals (i.e., signals from outside the DMA in which the satellite subscriber is located) have been reduced as follows:

- a satellite carrier may retransmit to a subscriber located in a community the signal of any station located outside the local market in which that subscriber is located *if* (1) the FCC had already determined, before the date of enactment of SHVERA, that the signal could be carried by a cable operator in that community because it was "significantly viewed" in that community, and such carriage was permissible under the FCC's network non-duplication and syndicated exclusivity rules; or (2) if the FCC determines, after the date of SHVERA enactment, that the signal is "significantly viewed" in the community in accordance with the same standards and procedures used to allow cable stations to carry "significantly viewed" signals.[52] In a Report and Order implementing SHVERA, the FCC identified thousands of instances in which broadcast signals met the "significantly viewed" criteria.[53]

- the retransmission (secondary transmission) of these "significantly viewed" broadcast signals are subject to a royalty-free compulsory copyright license — exempting satellite carriers from paying copyright royalties.[54]
- the FCC, in implementing SHVERA, changed its rules covering retransmission consent to allow a broadcaster located in a local market into which a satellite carrier is retransmitting the distant signal of a station that is affiliated with the same network as the local station to choose between making its signal available to satellite carriers based on retransmission consent (receiving compensation) vs. mandatory carriage (without compensation) on a county-specific basis rather than DMA-wide, as currently required.[55] This change was deemed necessary because satellite providers are now allowed to retransmit "significantly viewed" distant signals that may duplicate the network programming of a local station, but satellite carriage of those "significantly viewed" signals are determined on a county-by-county rather than DMA-wide basis, and thus it was felt that the local broadcaster should be able to make its retransmission consent/mandatory carriage election on a county-by- county basis.
- satellite carriers are allowed to retransmit the signal of WMUR, an ABC affiliate located in Manchester, New Hampshire, which is the only commercial station in that state affiliated with a major broadcast network, to any subscriber in that state,[56] subject to obtaining retransmission consent and meeting the provisions of the FCC's network non-duplication and syndication exclusivity rules. Such carriage is subject to royalty payments under the compulsory copyright license for the secondary transmission of distant broadcast signals.
- satellite carriers are allowed to retransmit the four commercial, network-affiliated stations that are located in the Burlington, Vermont DMA (which are the only commercial stations in that state) to any subscriber in either of the counties in that state outside the Burlington DMA (Windham and Bennington counties),[57] subject to obtaining retransmission consent and meeting the provisions of the FCC's network non-duplication and syndication

exclusivity rules. Such carriage is subject to royalty payments under the compulsory copyright license for the secondary transmission of distant broadcast signals.
- satellite carriers are allowed to retransmit the signals of any network- affiliated broadcast television station in Oregon to the four counties in Oregon (Umatilla, Grant, Malheur, and Wallowa) that are assigned to DMAs whose primary city is outside that state.[58] Also, a satellite carrier or cable company may elect to retransmit to subscribers in Umatilla, Grant, Malheur, and Wallowa counties in Oregon the broadcast signals of any television broadcast station in Oregon that any cable operator or satellite carrier was retransmitting to subscribers in those four counties on January 1, 2004.[59] These retransmissions to those four counties are subject to obtaining retransmission consent, to meeting the provisions of the FCC's network non-duplication and syndication exclusivity rules, and to royalty payments under the compulsory copyright license for the secondary transmission of distant broadcast signals.
- satellite carriers are allowed to retransmit the signals of all network-affiliated television broadcast station in Jackson, Mississippi to any subscriber in two counties (Wilkinson and Amite) in that state;[60] those counties are assigned to the Baton Rouge, Louisiana DMA. These retransmissions to those two counties are subject to obtaining retransmission consent, to meeting the provisions of the FCC's network non-duplication and syndication exclusivity rules, and to royalty payments under the compulsory copyright license for the secondary transmission of distant broadcast signals.
- the geographic areas in Alaska that are not in any Nielsen DMA are to be assigned by satellite carriers to one of the local markets (DMAs) in that state, in order to allow the carriers to offer subscribers in those areas the local-into-local service for the DMA to which they are assigned.[61]

In addition, satellite carriers with more than 5 million subscribers must retransmit all of the analog broadcast signals originating in Alaska and Hawaii within one year of the passage of SHVERA, and all of the digital broadcast signals originating in Alaska and Hawaii within 30 months of the passage of SHVERA. These signals must be made available to substantially all of the subscribers in their local markets (DMAs) and the signals from at least one of the local markets in the state must be made available to substantially all of the subscribers in the state not located in a DMA. The cost to subscribers of such transmission shall not exceed the cost of retransmission of local television stations in other states.[62]

Referring back to figure 3, SHVERA may expand upon the availability of programming of interest to subscriber Z if there are broadcast signals originating in state Y, but outside DMA F that meet the "significantly viewed" criteria, or if subscriber Z happens to be located in a state and county covered by one of the state-specific provisions in the act. Given the high likelihood that the satellite carrier already is uploading these broadcast signals to serve customers in the DMAs in which the signals originate (and the royalty-free compulsory copyright license for "significantly viewed" signals), it would appear that the only reasons that the satellite carrier might choose not to offer this additional programming to subscriber Z would be if it failed to negotiate a retransmission consent agreement with the license holder of the "significantly viewed" broadcast signal to cover subscriber Z (and other subscribers that previously could not be served) or if subscriber Z were not located in the footprint to which the signal was currently being beamed.

In some situations, the provisions in SHVERA that expand the number of instate signals potentially available to satellite subscribers will provide an additional public policy benefit. Candidates for public office who have had to reach many of the citizens of their state through high-priced advertising on out-of-state, big city stations may now be able to reach those citizens through lower-priced advertising on in-state stations. This could reduce the costs associated with political campaigns.

Chapter 5

ISSUES FOR CONGRESS

Localism remains one of the cornerstones of U.S. media policy. There are a small number of broadcast television stations relative to the number of local governmental jurisdictions. Moreover, every full power television station broadcasts signals that extend far beyond the borders of its city of license. Thus, when a particular station is assigned a city of license to serve, there will always be many nearby local jurisdictions that the licensee has no explicit or specific obligation to serve. Where the broadcast coverage area extends across governmental boundaries, and especially state borders, it is difficult for a broadcaster to fully address the needs of all jurisdictions. Broadcasters, of course, have the incentive to meet the needs and interests of as many of its potential viewers as possible. Most television broadcasters attempt to reconcile this by covering issues of general interest, such as crime and weather, and/or regional interest, such as transportation systems. However, some current statutory and regulatory requirements do not provide incentives, or even make it more difficult, for broadcast, cable, and satellite providers of television to meet the needs and interests of their communities. If Congress wants the FCC to systematically review its rules to eliminate any disincentives to localism or to clarify licensee obligations, it could pass legislation instructing the Commission to do so.

BROADCASTER OBLIGATIONS WITHIN THE CITY OF LICENSE

As explained earlier, the FCC's first priority when it assigns licenses is to provide general service to an area, and its second priority is to provide the first local service to a community. Most broadcast television stations are attentive to the needs and interests of the viewers in their city of license. It is in their self-interest to be responsive to their viewers. Their market incentives may diverge from this goal, however, if their city of license is an outlying city to a much larger city and their signal covers the larger city.

As shown in figure 4, the grade B contour of the station licensed to outlying city O fully covers major city M. In this situation, the licensee may have a stronger incentive to serve the needs and interests of the larger city. This incentive may be stronger yet if the city of license is in a different state than the larger city.

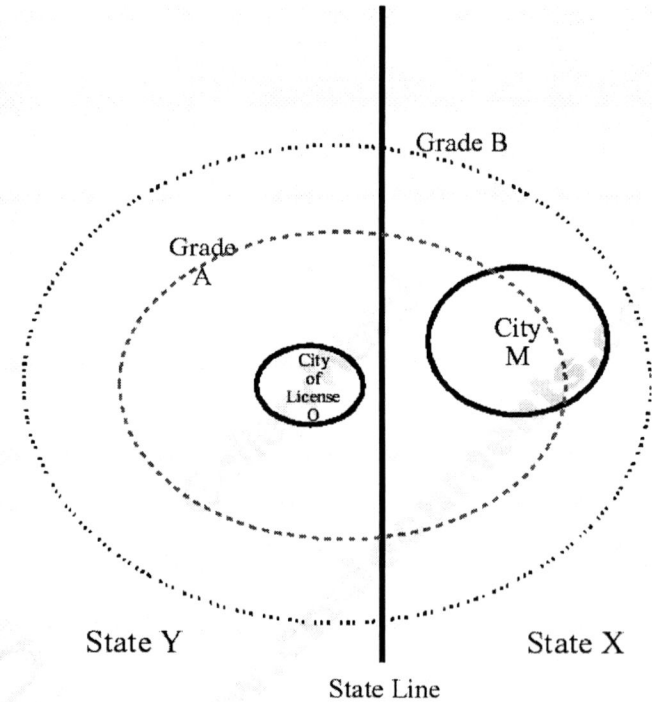

Source: CRS.

Figure 4. Broadcast station whose city of license is an outlying city to a major city, but whose signal covers the major city.

That incentive may affect how the station markets itself — for example, as a station in the outlying city of license or as a station in the larger city across the state line. This, in turn, may affect how the station is identified, geographically, by the industry. The *Television & Cable Factbook* is a widely used annual industry source book that presents data on each television station, by state. In the 2004 edition, seven television stations with city of license in one state, but located in a DMA whose principal city was in a neighboring state, were listed under the neighboring state. In the 2005 edition, five of those stations continued to be listed under the neighboring state; in the 2006 edition, none of them was listed under the neighboring state. Also, in the 2004 and 2005 editions, the city listings for all seven stations were hyphenated, with the large city listed first and the smaller city listed second. In the 2006 edition, each station was listed under the actual city of license only, with no reference to the larger city.[63] According to the publishers of the *Television & Cable Factbook*, its editors, not the stations, determined how to list each station. In 2006, it made the decision to list the stations by city (and, hence, state) of license, rather than using hyphenated market designations with the major city listed first.

During 2004, the FCC opened a proceeding on broadcast localism,[64] but to date has not proposed or adopted any rules relating to localism. Congress might choose to direct the FCC, in that proceeding, to undertake a rulemaking to explicitly identify, or provide written guidance about, the obligations of licensees with city of license in an outlying city to a major city to specifically serve the needs and interests of the viewers in their city of license.[65] It also might choose to direct the Commission to address how it would enforce those obligations. For example, under what circumstances, if any, could failure to serve the needs and interests of its community of license result in a license not being renewed (or being revoked)?

BROADCASTER OBLIGATIONS BEYOND THE CITY OF LICENSE

At the same time, current FCC rules are not clear about the broadcast television licensees' obligations to serve viewers within their service area who are beyond the borders of the city of license, but not in a larger nearby city with its own licensed broadcast television stations. As shown in figure 1, in many situations, the television household population beyond the city of license exceeds that within the city of license. This, in itself, provides broadcasters with some economic incentive to be responsive to the needs and interests of these viewers. But news and information programming is relatively expensive to produce, and unless such programming is of general interest to a relatively broad portion of the potential viewing audience, there is always the risk of losing audience. Thus, Congress might choose to direct the Commission, as part of its current proceeding on broadcast localism, to explicitly identify, or provide written guidance about, the obligations of licensees to serve the portion of their viewership that lies outside the city of license but not in large nearby cities with their own licensed broadcast television station. It also might choose to direct the Commission to address how it would enforce those obligations.

BROADCASTER OBLIGATIONS AND MULTICASTING

In 1999, the FCC issued a notice of inquiry concerning the public interest obligations of broadcast television licensees as they transition to digital television.[66] The Commission subsequently has issued two notices of proposed rulemaking as well as periodic reviews of the Commission's rules and policies affecting the conversion to digital television, and in September 2004 voted to adopt children's programming obligations for digital television broadcasters.[67] The Commission has incorporated the relevant portions of the comments received in those rulemakings and periodic reviews into its broadcast localism proceeding.[68]

Technological change has the potential to help broadcasters better meet the local needs of their viewers. With digital transmission, one option available to licensees is to use their 6 MHz of spectrum for multicasting — that is, to broadcast multiple programming streams. As the Commission develops rules addressing digital broadcast television public interest obligations, it might try to construct rules that foster programming that meets the possibly divergent needs of viewers within the city of license and viewers beyond the city of license. For example, it might consider modifying the current rule that requires cable operators to carry only the primary programming stream of each local television broadcaster by requiring cable operators to carry each programming stream that offers distinct programming aimed at a different, previously unserved geographic portion of the broadcaster's serving area.[69] This could explicitly address those situations in which a broadcaster's serving area crosses state borders, awarding the broadcaster must carry rights for a second signal if the programming on that signal specifically addresses the needs and interests of the viewing households in the second state. If the FCC were to consider this approach, it would want to take into account the impact on cable systems of requiring them to carry additional broadcast channels. It also would want to determine how best to construct a rule that did not artificially encourage or discourage broadcasters from choosing multicasting over other potential applications of digital technology to their 6 MHz of spectrum, such as high definition television. Congress might choose to direct the FCC, in its current proceeding on implementation of the digital transition, to study and construct recommendations for rules (and, if necessary, statutory changes) to address the potentially related issues of mandatory carriage of multiple broadcast signals and better serving the needs and interests of viewers in different governmental jurisdictions.

INCREASING THE FLEXIBILITY OF CABLE CARRIAGE RULES

As explained earlier, the existing array of must carry and non-duplication rules and compulsory copyright license fees may restrict or discourage cable operators that happen to be located in a DMA that has its primary city in another state from carrying the signals of broadcasters in their own state that are located in a different DMA. This can decrease viewer access to both informational and entertainment programming of state-wide interest. The data presented in table 1 suggest this may not be an isolated occurrence. In many states a substantial number and percentage of television households are in DMAs in which the primary city is located outside the state and in which most of the television stations have city of license outside the state.

However, there is a degree of flexibility in the must carry rules (the statutory provision allowing cable operators to request that they be allowed to carry signals from outside their DMA that would foster localism), the non-duplication rules (allowing "significantly viewed" stations to be carried without having duplicated programming blocked), and the copyright laws (providing a royalty-free permanent compulsory copyright license for the secondary transmission of programming of broadcast stations that are "significantly viewed" in the cable system's service area).

Congress might choose to direct the FCC, when reviewing its existing rules as part of its current broadcast localism proceeding, to heed the flexibility that Congress has given it to implement and administer its rules in a fashion that fosters localism. It might instruct the Commission to study whether there are narrowly-defined conditions under which the existing non-duplication rules can be loosened to foster the cable carriage of programming of state-wide interest without undermining the goals and objectives of those rules. It also might ask the Commission to explore how it could best allow exceptions to its current rule that uses DMAs to determine which broadcast television signals a cable company must carry. As indicated earlier, the Commission has ruled that only broadcast licensees and cable operators have standing to request exceptions to the current rule restricting must carry rights to stations within the DMA. The Commission might investigate whether it would be in the public interest for other parties, such as state officials, to be able to make such a request based on a demonstratively positive impact on localism. It also might investigate whether it would be in the public interest for the Commission, itself, to have the right to propose an exception to the rule on its own authority. If it were to reach the conclusion that such authority would be in the public interest, it might recommend to Congress that the statute be modified to give it that authority.[70]

The Commission already has concluded that the parties currently with standing to seek an exception — the broadcast licensee and the cable system operator — have knowledge of key relevant parameters (for example, the demand for particular types of programming, the programming available both on the specific broadcast station and on the cable system, the geographic reach of the broadcast station's grade B contours, etc.) not readily available to state officials or the FCC. Adding or deleting must carry stations will change the array of programming available to the cable system's subscribers and any party seeking to change the line-up of channels should have sufficient information on subscribers' preferences to be confident that consumers will be better served by the proposed change in programming.

INCREASING THE FLEXIBILITY OF SATELLITE LOCAL-INTO-LOCAL PROGRAMMING

By passing SHVERA in November 2004, Congress expanded the scope of instate television programming that satellite operators are permitted (but not required) to offer subscribers. In Alaska and Hawaii, that expansion in mandatory. The combination of a generic change in law — allowing satellite providers to offer programming that is "significantly viewed" by over-the-air television viewers — and several state-specific provisions intended to address restricted access to programming in six states may significantly reduce consumer complaints that they are not able to receive programming via satellite that meets their needs and interests. The effectiveness of SHVERA, however, will not be determined until it has been in operation long enough to find out if the criteria in the rules and limitations associated with the "significantly viewed" provision allow for a real expansion in the signals made available to satellite subscribers. If the "significantly viewed" provision does not provide relief for subscribers seeking programming that meets state-specific needs and interests, then it is likely that bills will be introduced in the 110th Congress that seek state-specific or county-specific solutions analogous to the ones involving New Hampshire, Vermont, Mississippi, Oregon, Alaska, and Hawaii in SHVERA. Several such bills already have been introduced, as described below.

BILLS INTRODUCED IN THE 110TH CONGRESS

To date, three bills that address cable and satellite carriage of local broadcast television station signals have been introduced in the 110th Congress. Senator Allard has introduced S. 124, which would allow satellite operators to offer subscribers located in two counties in the southwestern corner of Colorado the signals of broadcasters in Denver, even though those counties are not located in the Denver DMA (analogous to the provisions in SHVERA affecting subscribers in certain counties in New Hampshire, Vermont, Mississippi, and Oregon), and also would allow cable operators in those two counties to carry the primary signal of any network station located in Denver. Senator Salazar has introduced S. 760, which, in addition to the two provisions in S. 124, would waive the retransmission rules to allow a satellite carrier, cable system, or translator station to carry the primary signal of a network station located in a state to subscribers in that state who otherwise would not receive the primary signal of that network because those subscribers are located in a DMA outside of the state if two conditions are met: (1) the FCC determines that it is in the best interest of the public welfare, and (2) the satellite carrier, cable system, or translator station agrees to also carry the primary signal of the network station in the assigned DMA.

Representative Boren has introduced H.R. 602, which would modify the retransmission rules to allow a satellite carrier to provide the signals of network stations located in Oklahoma to subscribers who reside in Oklahoma but do not currently receive the signal of any network station located in that state because of their assignment to a DMA receiving network stations located outside Oklahoma, if those subscribers choose to receive the Oklahoma signals rather than the out-of-state signals.

Representative Ross has introduced H.R. 2821, which would: amend section 352(b)(2) of the Communictions Act (47 U.S.C. § 352(b)(2)) to permit satellite carriers and cable operators to retransmit the signals of local television broadcast stations to any DMA that is adjacent to, and at least partially located in the same state as, the DMA in which the broadcast station is located; amend section 122 of the U.S. Copyright Act (17 U.S.C. § 122) to allow satellite operators to retransmit the signals of those local broadcast stations into those adjacent DMAs under a royalty-free statutory copyright license; and instruct the FCC to revise the regulations concerning network non-duplication protection, syndicated exclusivity protection, and sports blackout protection (47 CFR § 76) to permit retransmission if the subscriber receiving the signals is located in any of those adjacent DMAs.

Table 1. Television Households in Each State that Are Located in Designated Market Areas (DMAs) for which the Primary City is Outside the State

State	# TV households	Counties in DMAs for which primary city is outside the state DMA: County	# TV households in DMAs for which primary city is outside the state	% of TV households in DMAs for which primary city is outside the state	Currently operating full power broadcast TV stations in DMAs for which primary city is outside the state
Alabama	1,768,300	Atlanta, GA DMA: Cleburne, Randolph	5,790+8,960+	7.16%	Atlanta, GA DMA: no station with city of license in AL;
		Columbus, GA DMA: Chambers, Lee, Russell, Barbour	14,660+48,700+19,770+1 0 ,650+		Columbus, GA DMA: 1 (UHF analog and digital) commercial station with city of license in Opelika, AL and 1 (UHF analog and digital) non-commercial station transmitting from Louisville, AL;
		Columbus-Tupelo-West Point, MS DMA: Lamar	6,450+		Columbus-Tupelo-West Point, MS DMA: no station with city of license in AL;
		Meridian, MS DMA: Sumter, Choctaw	5,500+6,190= 126,670		Meridian, MS DMA: no station with city of license in AL
Alaska	193,630	none — but some extremely low density areas lie outside DMAs	0	0.00%	

Table 1. (Continued)

State	# TV households	Counties in DMAs for which primary city is outside the state DMA: County	# TV households in DMAs for which primary city is outside the state	% of TV households in DMAs for which primary city is outside the state	Currently operating full power broadcast TV stations in DMAs for which primary city is outside the state
Arizona	2,046,350	Albuquerque-Santa Fe, NM DMA: Apache-North	13,390	0.65%	Albuquerque-Santa Fe, NM DMA: no station with city of license in AZ
Arkansas	1,057,360	Memphis, TN DMA: Mississippi, Crittenden, Poinsett, Cross, Saint Francis, Lee, Phillips	18,740+18,660+ 9,930+7,380+ 9,780+ 4,070+ 9,020+	21.17%	Memphis, TN DMA: no station with city of license in AR;
		Springfield, MO DMA: Fulton, Baxter, Marion, Boone, Newton, Carroll	4,790+ 16,900+6,940+ 14,220+3,530+ 10,380+		Springfield, MO DMA: 1 UHF analog commercial station with city of license in Eureka Springs, AR and 1 UHF analog commercial station with city of license in Harrison, AR;

State	# TV households	Counties in DMAs for which primary city is outside the state DMA: County	# TV households in DMAs for which primary city is outside the state	% of TV households in DMAs for which primary city is outside the state	Currently operating full power broadcast TV stations in DMAs for which primary city is outside the state
		Shreveport, LA DMA: Howard, Sevier, Little River, Hempstead, Nevada, Miller, Lafayette, Columbia,	5,370+5,560+5,370+ 8,840+3,790+15,650+ 3,330+9,680+		Shreveport, LA DMA: no station with city of license in AR;
		Greenwood-Greenville, MS DMA: Chicot	4,950+		Greenwood-Greenville, MS DMA: no station with city of license in AR;
		Monroe, LA-El Dorado, AR DMA: Union, Ashley	17,760+ 9,230= 223,870		Monroe, LA.-El Dorado, AR DMA: no station with city of license in AR
California	11,774,780	Reno, NV DMA: Alpine, El Dorado East, Mono, Lassen	480+13,770+4,690+ 9,560+	0.74%	Reno, NV DMA: no station with city of license in CA;
		Medford-Klamath Falls, OR DMA: Siskiyou	17,610+		Medford-Klamath Falls, OR DMA: no station with city of license in CA;

Table 1. (Continued)

State	# TV households	Counties in DMAs for which primary city is outside the state DMA: County	# TV households in DMAs for which primary city is outside the state	% of TV households in DMAs for which primary city is outside the state	Currently operating full power broadcast TV stations in DMAs for which primary city is outside the state
		Yuma, AZ-El Centro, CA DMA: Imperial	41,550= 87,660		Yuma, AZ-El Centro, CA DMA: 1 VHF analog Fox-affiliated commercial station with city of license in El Centro, CA and 1 VHF analog Univision-affiliated commercial station with city of license in El Centro
Colorado	1,738,830	Albuquerque, NM DMA: Montezuma, La Plata	9,390+17,340=26,730	1.54%	Albuquerque, NM DMA: 1 (VHF analog and digital) CBS-affiliated commercial station that is a satellite of an Albuquerque station, and 1 UHF analog Telemundo-affiliated commercial station that is a satellite of an Albuquerque station, all with city of license in Durango, CO

State	# TV households	Counties in DMAs for which primary city is outside the state DMA: County	# TV households in DMAs for which primary city is outside the state	% of TV households in DMAs for which primary city is outside the state	Currently operating full power broadcast TV stations in DMAs for which primary city is outside the state
Connecticut	1,331,810	New York City, NY DMA: Fairfield	330,490	24.82%	New York City, NY DMA: 1 UHF analog commercial station with city of license in Bridgeport, CT, and 1 (UHF analog and digital) non-commercial station transmitting from Bridgeport, CT
Delaware	313,630	Philadelphia, PA DMA: Kent, New Castle	49,460+195,540+	100.00%	Philadelphia, PA DMA: 1 UHF analog commercial station with city of license in Wilmington, DE, and 1 (UHF analog and digital) non-commercial station transmitting from Wilmington;
		Salisbury, MD DMA: Sussex	68,630= 313,630		Salisbury, MD DMA: 1 UHF analog noncommercial station transmitting from Seaford, DE
District of Columbia	244,270	none	0	0.00%	

Table 1. (Continued)

State	# TV households	Counties in DMAs for which primary city is outside the state DMA: County	# TV households in DMAs for which primary city is outside the state	% of TV households in DMAs for which primary city is outside the state	Currently operating full power broadcast TV stations in DMAs for which primary city is outside the state
Florida	6,728,860	Mobile, AL-Pensacola-Fort Walton Beach, FL DMA: Okaloosa, Santa Rosa, Escambia	71,260+47,830+115,610= 234,700	3.49%	Mobile, AL-Pensacola-Fort Walton Beach, FL DMA: 3 UHF analog commercial stations with city of license in Fort Walton Beach, FL, 3 (UHF analog and digital) commercial stations (including 1 ABC affiliate) with city of license in Pensacola, FL, and 1 (UHF analog and digital) noncommercial station transmitting from Pensacola, FL
Georgia	3,195,950	Greenville-Spartanburg-Anderson, SCAsheville, NC DMA: Stephens, Franklin, Hart, Elbert	10,390+8,290+9,650+8,320+	9.58%	Greenville-Spartanburg-Anderson, SCAsheville, NC DMA: 1 UHF analog CBSaffiliated commercial station with city of license in Toccoa, GA;

State	# TV households	Counties in DMAs for which primary city is outside the state DMA: County	# TV households in DMAs for which primary city is outside the state	% of TV households in DMAs for which primary city is outside the state	Currently operating full power broadcast TV stations in DMAs for which primary city is outside the state
		Jacksonville, FL DMA: Charlton, Camden, Ware, Glynn, Brantley, Pierce	3,400+ 15,490+13,310+ 28,130+5,930+6,250+		Jacksonville, FL DMA: 1 UHF analog commercial station with city of license in Brunswick, GA;
		Chattanooga, TN DMA: Dade, Walker, Catoosa, Whitfield, Murray, Chatooga	5,980+24,050+22,230+ 30,010+14,780+ 10,260+		Chattanooga, TN DMA: 1 UHF analog commercial station with city of license in Dalton, GA, and 1 (UHF analog and digital) noncommercial station transmitting from Chatsworth-Dalton, GA;
		Dothan, AL DMA: Early, Seminole	4,820+3,640+		Dothan, AL DMA: no station with city of license in GA;
		Tallahassee, FL-Thomasville, GA DMA: Decatur, Grady, Thomas, Brooks, Lowndes, Lanier, Echols	10,680+9,080+16,940+6, 430+33,980+2,730+ 1,360= 306,130		Tallahassee, FL-Thomasville, GA DMA: 1 UHF analog FOX-affiliated commercial station with city of license in Bainbridge, GA, and 1 VHF analog CBS-affiliated commercial station with city of license in Thomasville, GA

Table 1. (Continued)

State	# TV households	Counties in DMAs for which primary city is outside the state DMA: County	# TV households in DMAs for which primary city is outside the state	% of TV households in DMAs for which primary city is outside the state	Currently operating full power broadcast TV stations in DMAs for which primary city is outside the state
Hawaii	412,190	none	0	0.00%	
ID	486,450	Salt Lake City, UT DMA: Oneida, Franklin, Bear Lake	1,430+3,650+ 2,340+	24.02%	Salt Lake City, UT DMA: no station with city of license in ID;
		Spokane, WA DMA: Boundary, Bonner, Shoshone, Kootenai, Benewah, Latah, Idaho, Clearwater, Lewis, Nez Perce	3,450+ 14,710+ 5,760+ 43,920+3,560+ 12,330+5,840+3,390+ 1,510+14,950= 116,840		Spokane, WA DMA: 1 (VHF analog and digital) CBS-affiliated commercial station, affiliated with a station in Yakima, WA, with city of license in Lewiston, ID, 1 (UHF analog and digital) noncommercial station transmitting from Couer d'Alene, ID, and 1 (UHF analog and digital) noncommercial station transmitting from Moscow, ID

State	# TV households	Counties in DMAs for which primary city is outside the state DMA: County	# TV households in DMAs for which primary city is outside the state	% of TV households in DMAs for which primary city is outside the state	Currently operating full power broadcast TV stations in DMAs for which primary city is outside the state
Illinois	4,648,990	St. Louis, MO DMA: Randolph, Monroe, St. Clair, Washington, Clinton, Marion, Clay, Fayette, Montgomery, Macoupin, Greene, Jersey, Calhoun, Bond, Madison	12,170+11,020+96,610+5,830+13,050+16,190+5,770+8,160+11,380+19,590+5,560+8,2 20+2,060+6,340+103,330+	14.77%	St. Louis, MO DMA: 1 (UHF analog and digital) commercial station with city of license in East St. Louis, IL;
		Evansville, IN DMA: Wayne, Edwards, Wabash, White	7,180+2,890+5,140+6,430+		Evansville, IN DMA: no station with city of license in IL;
		Terre Haute, IN DMA: Clark, Jasper, Crawford, Richland, Lawrence	7,150+3,850+7,750+6,560+6,180+		Terre Haute, IN DMA: 1 (UHF analog and digital) noncommercial station transmitting from Olney, IL;

Table 1. (Continued)

State	# TV households	Counties in DMAs for which primary city is outside the state DMA: County	# TV households in DMAs for which primary city is outside the state	% of TV households in DMAs for which primary city is outside the state	Currently operating full power broadcast TV stations in DMAs for which primary city is outside the state
		Paducah, KY-Cape Girardeau, MO-Mount Vernon, IL DMA: Jefferson, Perry, Franklin, Hamilton, Gallatin, Saline, Williamson, Jackson, Union, Johnson, Hardin, Pope, Massac, Pulaski, Alexander	15,560+8,940+16,240+ 3,330+2,650+10,810+ 25,760+23,450+7,320+ 4,310+1,990+1,770+ 6,210+2,860+3,720+		Paducah, KY-Cape Girardeau, MO-Mount Vernon, IL DMA: 1 (UHF analog and digital) commercial station with city of license in Marion, IL, 1 (VHF analog and digital) ABCaffiliated commercial station with city of license in Harrisburg, IL, 1 VHF analog commercial station with city of license in Mt. Vernon, IL, 1 (VHF analog and digital) noncommercial station transmitting from Carbondale, IL;
		Davenport, IA-Rock Island-Moline, IL DMA: Henderson, Warren, Knox, Mercer, Henry,			

State	# TV households	Counties in DMAs for which primary city is outside the state DMA: County	# TV households in DMAs for which primary city is outside the state	% of TV households in DMAs for which primary city is outside the state	Currently operating full power broadcast TV stations in DMAs for which primary city is outside the state
Indiana	2,382,900	Chicago, IL DMA: Lake, Porter, La Porte, Newton, Jasper	182,170+56,790+41,320+5,460+11,160+	21.90%	Chicago, IL DMA: 1 (UHF analog and digital) commercial station with city of license in Hammond, IN, and 1 UHF analog noncommercial station transmitting from Gary, IN;
		Cincinnati, OH DMA: Union, Franklin, Ripley, Dearborn, Ohio, Switzerland	2,760+7,980+10,270+17,540+2,270+3,670+		Cincinnati, OH DMA: no station with city of license in IN;
		Louisville, KY DMA: Harrison, Floyd, Clark, Crawford, Orange, Washington, Scott, Jefferson, Jennings, Jackson	13,660+27,810+40,120+4, 260+7,570+10,540+ 9,120+12,330+10,610+ 16,460+		Louisville, KY DMA: 1 (UHF analog and digital) commercial station with city of license in Salem, IN;
		Dayton, OH DMA: Wayne	27,950= 521,820		Dayton, OH DMA: 1 (UHF analog and digital) commercial station with city of license in Richmond, IN

Table 1. (Continued)

State	# TV households	Counties in DMAs for which primary city is outside the state DMA: County	# TV households in DMAs for which primary city is outside the state	% of TV households in DMAs for which primary city is outside the state	Currently operating full power broadcast TV stations in DMAs for which primary city is outside the state
Iowa	1,152,630	Omaha, NE DMA: Crawford, Harrison, Shelby, Cass, Pottawattamie, Mills, Montgomery, Fremont, Page	6,360+6,270+5,150+ 6,130+34,260+5,470+ 4,670+3,160+ 6,510+	12.68%	Omaha, NE DMA: 1 (UHF analog and digital) noncommercial station transmitting from Council Bluffs, IA, and 1 UHF analog noncommercial station transmitting from Red Oak, IA;
		Sioux Falls, SD DMA: Lyon, Osceola,	4,210+2,750+		Sioux Falls, SD DMA: no station with city of license in IA;
		Rochester, MN-Mason City, IA-Austin, MN DMA: Winnebago, Worth, Mitchell, Howard, Hancock, Cerro Gordo, Floyd	4,670+3,280+4,130+ 3,850+4,740+19,070+ 6,730+		Rochester, MN-Mason City, IA-Austin, MN DMA: 1 (VHF analog and digital) CBS-affiliated commercial station with city of license in Mason City, IA, and 1 UHF analog noncommercial station transmitting from Mason City, IA;

State	# TV households	Counties in DMAs for which primary city is outside the state DMA: County	# TV households in DMAs for which primary city is outside the state	% of TV households in DMAs for which primary city is outside the state	Currently operating full power broadcast TV stations in DMAs for which primary city is outside the state
		Quincy, IL-Hannibal, MO-Keokuk, IA DMA: Lee	4,770= 146,180		Quincy, IL-Hannibal, MO-Keokuk, IA DMA: no station with city of license in IA
Kansas	1,044,100	Kansas City, MO DMA: Atchison, Leavenworth, Wyandotte, Douglas, Johnson, Franklin, Miami, Anderson, Linn	6,240+23,870+ 58,030+38,970+ 186,740+10,910+9,510+3 150+3,950+	40.76%	Kansas City, MO DMA: 1 (UHF analog and digital) commercial station with city of license in Lawrence, KS;
		Tulsa, OK DMA: Chautaqua, Montgomery	1,750+14,210+		Tulsa, OK DMA: no station with city of license in KS;
		Lincoln and Hastings-Kearney, NE DMA: Phillips, Smith, Jewell, Republic	2,350+1,880+1,690+ 2,480+		Lincoln and Hastings-Kearney, NE DMA: no station with city of license in KS;
		St. Joseph, MO DMA: Doniphan	3,290+		St. Joseph, MO DMA: no station with city of license in KS;

Table 1. (Continued)

State	# TV households	Counties in DMAs for which primary city is outside the state DMA : County	# TV households in DMAs for which primary city is outside the state	% of TV households in DMAs for which primary city is outside the state	Currently operating full power broadcast TV stations in DMAs for which primary city is outside the state
		Joplin, MO-Pittsburg, KS DMA: Woodson, Allen, Bourbon, Wilson, Neosho, Crawford, Labette, Cherokee	1,540+5,610+6,090+ 3,850+6,640+15,180+ 8,890+8,770= 425,590		Joplin, MO-Pittsburg, KS DMA: 1 (VHF analog and digital) CBS-affiliated commercial station with city of license in Pittsburg, KS
Kentucky	1,624,650	Nashville, TN DMA: Trigg, Christian, Todd, Logan, Simpson, Allen, Monroe, Cumberland, Clinton	5,380+24,600+4,310+ 10,340+6,500+7,090+ 4,700+2,880+4,080+	31.33%	Nashville, TN DMA: no station with city of license in KY;
		Cincinnati, OH DMA: Kenton, Campbell, Gallatin, Owen, Grant, Pendleton, Bracken, Mason, Roberston, Boone	60,400+35,400+2,960+ 4,080+8,890+5,360+ 3,280+6,850+870+ 35,080+		Cincinnati, OH DMA: 1 (UHF analog and digital) noncommercial station transmitting from Covington, KY, and 1 (UHF analog and digital) noncommercial station transmitting from Owenton, KY;

State	# TV households	Counties in DMAs for which primary city is outside the state DMA: County	# TV households in DMAs for which primary city is outside the state	% of TV households in DMAs for which primary city is outside the state	Currently operating full power broadcast TV stations in DMAs for which primary city is outside the state
		Knoxville, TN DMA: McCreary, Bell, Harlan,	6,520+11,980+12,820+		Knoxville, TN DMA: 1 UHF analog commercial station with city of license in Harlan, KY;
		Charleston-Huntington, WV DMA: Lewis, Greenup, Carter, Boyd, Elliott, Lawrence, Johnson, Martin, Floyd, Pike	5,550+14,660+10,590+19, 650+2,760+6,070+ 8,880+4,920+16,960+ 27,020+		Charleston-Huntington, WV DMA: 1 (UHF analog and digital) commercial station with city of license in Ashland, KY; 1 (UHF analog and digital) noncommercial station transmitting from Ashland, KY; 1 (UHF analog and digital) noncommercial station transmitting from Pikeville, KY;
		Tri-Cities (Kingsport-Johnson City, TN/Bristol, VA) DMA: Letcher, Leslie	9,930+4,680+		Tri-Cities (Kingsport-Johnson City, TN/Bristol, VA) DMA: no station with city of license in KY;
Louisiana	1,667,710	none	0	0.00%	
Maine	534,740	none	0	0.00%	

Table 1. (Continued)

State	# TV households	Counties in DMAs for which primary city is outside the state DMA: County	# TV households in DMAs for which primary city is outside the state	% of TV households in DMAs for which primary city is outside the state	Currently operating full power broadcast TV stations in DMAs for which primary city is outside the state
Maryland	2,075,720	Washington, DC DMA: Frederick, Washington, Montgomery, Prince George's Charles, Calvert, St. Mary's, Allegheny	77,010+51,710+341,700+ 298,970+ 45,330+28,310+32,390+2 9,350+	44.13%	Washington, DC DMA: 1 (UHF analog and digital) commercial NBC-affiliated station with city of license in Hagerstown, MD, 1 UHF analog commercial station with city of license in Hagerstown, MD, 1 (UHF analog and digital) noncommercial station transmitting from Hagerstown, MD, and 1 UHF analog noncommercial station transmitting from Frederick, MD
		Pittsburgh, PA DMA: Garrett	11,270=916,040		Pittsburgh, PA DMA: 1 UHF analog noncommercial station transmitting from Oakland, MD

State	# TV households	Counties in DMAs for which primary city is outside the state DMA: County	# TV households in DMAs for which primary city is outside the state	% of TV households in DMAs for which primary city is outside the state	Currently operating full power broadcast TV stations in DMAs for which primary city is outside the state
Massachusetts	2,487,160	Albany-Schenectady-Troy, NY DMA: Berkshire	55,730+	10.76%	Albany-Schenectady-Troy, NY DMA: 1 (UHF analog and digital) ABC-affiliated commercial station that is a satellite of an Albany station with city of license in Adams, MA
		Providence, RI-New Bedford, MA DMA: Bristol	211,920= 267,650		Providence, RI-New Bedford, MA DMA: no station with city of license in MA
Michigan	3,867,220	Green Bay-Appleton, WI DMA: Menominee	10,760+	3.60%	Green Bay-Appleton, WI DMA: no station with city of license in MI (1 VHF analog CBS-affiliated commercial station with city of license in Escanaba, MI is in the Marquette, MI DMA but is a satellite of a Green Bay station);
		Toledo, OH DMA: Lenawee	37,380+		Toledo, OH DMA: no station with city of license in MI;

Table 1. (Continued)

State	# TV households	Counties in DMAs for which primary city is outside the state DMA: County	# TV households in DMAs for which primary city is outside the state	% of TV households in DMAs for which primary city is outside the state	Currently operating full power broadcast TV stations in DMAs for which primary city is outside the state
		South Bend-Elkhart, IN DMA: Berrien, Cass	63,850+20,120+		South Bend-Elkhart, IN DMA: no station with city of license in MI;
		Duluth, MN-Superior, WI DMA: Gogebic	7,060= 139,170		Duluth, MN-Superior, WI DMA: no station with city of license in MI
Minnesota	1,951,070	Sioux Falls, SD DMA: Lincoln, Pipestone, Murray, Rock, Nobles	2,640+3,940+ 3,670+3,960+7,910+	7.59%	Sioux Falls, SD DMA: 1 UHF analog noncommercial station transmitting from Worthington, MN;

State	# TV households	Counties in DMAs for which primary city is outside the state DMA: County	# TV households in DMAs for which primary city is outside the state	% of TV households in DMAs for which primary city is outside the state	Currently operating full power broadcast TV stations in DMAs for which primary city is outside the state
		Fargo-Valley City, ND DMA: Kittson, Roseau, Lake of the Woods, Marshall, Pennington, Red Lake, Polk, Clearwater, Norman, Mahnomen, Clay, Becker, Wilkin, Otter Tail	2,070+6,300+ 1,820+4,040+5,530+ 1,660+12,480+3,220+ 2,960+1,980+19,260+ 12,130+2,690+22,990+		Fargo-Valley City, ND DMA: 1 VHF analog FOX-affiliated station that is a satellite of a Fargo station with city of license in Thief River Falls, MN;
		La Crosse-Eau Claire, WI DMA: Winona, Houston	19,090+7,820= 148,160		La Crosse-Eau Claire, WI DMA: no station with city of license in MN
Mississippi	1,059,080	New Orleans, LA DMA: Pearl River, Hancock	18,960+18,040+	17.60%	New Orleans, LA DMA: no station with city of license in MS;
		Memphis, TN DMA: De Soto, Tunica, Coahoma, Quitman, Panola, Tate, Lafayette, Marshall, Benton, Tippah, Alcorn	44,080+3,510+9,880+ 3,400+12,680+9,270+ 15,380+12,520+2,970+ 8,130+14,500+		Memphis, TN DMA: 1 (UHF analog and digital) commercial station with city of license in Holly Springs, MS and 1 UHF analog noncommercial station transmitting from Oxford, MS;

Table 1. (Continued)

State	# TV households	Counties in DMAs for which primary city is outside the state DMA: County	# TV households in DMAs for which primary city is outside the state	% of TV households in DMAs for which primary city is outside the state	Currently operating full power broadcast TV stations in DMAs for which primary city is outside the state
		Mobile, AL-Pensacola-Fort Walton Beach, FL DMA: Greene	4,220+		Mobile, AL-Pensacola-Fort Walton Beach, FL DMA: no station with city of license in MS;
		Baton Rouge, LA DMA: Wilkinson, Amite	3,590+5,270= 186,400		Baton Rouge, LA DMA: no station with city of license in MS
Missouri	2,233,240	Omaha, NE DMA: Atchison	2,570+	8.64%	Omaha, NE DMA: no station with city of license in MO;
		Jonesboro, AR DMA: Ripley	5,250+		Jonesboro, AR DMA: no station with city of license in MO;

State	# TV households	Counties in DMAs for which primary city is outside the state DMA: County	# TV households in DMAs for which primary city is outside the state	% of TV households in DMAs for which primary city is outside the state	Currently operating full power broadcast TV stations in DMAs for which primary city is outside the state
		Paducah, KY-Cape Girardeau-Harrisburg, MO-Mount Vernon, IL DMA: Perry, Madison, Bollinger, Cape Girardeau, Carter, Wayne, Butler, Stoddard, Scott, Mississippi, New Madrid, Dunklin, Pemiscot	7,020+4,710+4,700+27,510+2,340+5,730+16,790+12,120+15,770+5,160+7,640+13,100+7,670+		Paducah, KY-Cape Girardeau-Harrisburg, MOMount Vernon, IL DMA: 1 (UHF analog and digital) FOX-affiliated commercial station with city of license in Cape Girardeau, MO, 1 (VHF analog and digital) CBS-affiliated commercial station with city of license in Cape Girardeau, and 1 UHF analog commercial station that is a satellite of a Harrisburg, IL station with city of license in Poplar Bluff, MO;
		Quincy, IL-Hannibal, MO-Keokuk, IA DMA: Clark, Lewis, Marion, Ralls, Monroe, Shelby, Knox,	2,950+4,040+11,000+3,880+3,960+2,760+1,760+		Quincy, IL-Hannibal, MO-Keokuk, IA DMA: 1 (VHF analog and digital) CBS-affiliated commercial station with city of license in Hannibal, IL

Table 1. (Continued)

State	# TV households	Counties in DMAs for which primary city is outside the state DMA: County	# TV households in DMAs for which primary city is outside the state	% of TV households in DMAs for which primary city is outside the state	Currently operating full power broadcast TV stations in DMAs for which primary city is outside the state
		Ottumwa, IA-Kirksville, MO DMA: Putnam, Schuyler, Scotland, Sullivan, Adair, Macon	2,260+1,670+1,830+ 2,860+9,540+6,360= 192,950		Ottumwa, IA-Kirksville, MO DMA: no station with city of license in MO
Montana	354,900	Spokane, WA DMA: Lincoln	7,370+	5.24%	Spokane, WA DMA: no station with city of license in MT;
		Minot-Bismarck-Dickinson, ND DMA: Sheridan, Daniels, Roosevelt, Richland, McCone, Wibaux	1,520+890+3,420+ 3,750+780+390+		Minot-Bismarck-Dickinson, ND DMA: no station with city of license in MT;
		Rapid City, SD DMA: Carter	480= 18,600		Rapid City, SD DMA: no station with city of license in MT

State	# TV households	Counties in DMAs for which primary city is outside the state DMA: County	# TV households in DMAs for which primary city is outside the state	% of TV households in DMAs for which primary city is outside the state	Currently operating full power broadcast TV stations in DMAs for which primary city is outside the state
Nebraska	675,030	Denver, CO DMA: Kimball, Cheyenne, Deuel, Keith, Garden, Grant, Hooker, Dawes, Box Butte	1,690+4,070+890+3,620+980+290+290+3,380+4,670+	12.53%	Denver, CO DMA: 1 (VHF analog and digital) noncommercial station transmitting from Alliance, NE;
		Wichita-Hutchinson, KS DMA: Dundy	860+		Wichita-Hutchinson, KS DMA: no station with city of license in NE:
		Sioux Falls, SD DMA: Cherry	2,420+		Sioux Falls, SD DMA: 1 (VHF analog and digital) noncommercial station transmitting from Merriman, NE;
		Sioux City, IA DMA: Dakota, Thurston, Dixon, Cedar, Wayne, Stanton, Knox, Pierce, Madison	7,270+2,190+2,380+3,580+3,380+2,280+3,700+2,880+13,450+		Sioux City, IA DMA: 1 (UHF analog and digital) noncommercial station transmitting from Norfolk, NE;
		Rapid City, SD DMA: Sioux, Sheridan, Morrill, Banner	590+2,460+2,070+300+		Rapid City, SD DMA: no station with city of license in NE;

Table 1. (Continued)

State	# TV households	Counties in DMAs for which primary city is outside the state DMA: County	# TV households in DMAs for which primary city is outside the state	% of TV households in DMAs for which primary city is outside the state	Currently operating full power broadcast TV stations in DMAs for which primary city is outside the state
		Cheyenne, WY-Scottsbluff, NE DMA: Scottsbluff	14,870= 84,560		Cheyenne, WY-Scottsbluff, NE DMA: 1 (VHF analog and digital) ABC- affiliated commercial station with city of license in Hay Springs, NE and 1 (VHF analog and digital) CBS-affiliated commercial station that is a satellite of a Cheyenne station with city of license in Scottsbluff, NE
Nevada	833,960	Salt Lake City, UT DMA: Elko, Eureka, White Pine	15,390+560+2,730= 18,680	2.24%	Salt Lake City, UT DMA: 1 VHF analog NBC-affiliated commercial station with city of license in Elko
New Hampshire	498,150	Portland-Auburn, ME DMA: Coos, Carroll	14,090+19,150+	100.00%	Portland-Auburn, ME DMA: no station with city of license in NH;

State	# TV households	Counties in DMAs for which primary city is outside the state DMA: County	# TV households in DMAs for which primary city is outside the state	% of TV households in DMAs for which primary city is outside the state	Currently operating full power broadcast TV stations in DMAs for which primary city is outside the state
		Burlington, VT-Plattsburgh, NY DMA: Grafton, Sullivan	32,570+16,910+		Burlington, VT-Plattsburgh, NY DMA: 1 (UHF analog and digital) non-commercial station transmitting from Littleton, NH;
		Boston, MA-Manchester, NH DMA: Cheshire, Hillsborough, Belknap, Merrimack, Strafford, Rockingham	28,940+151,430+24,310+55,020+44,910+110,820= 498,150		Boston, MA-Manchester, NH DMA: 1 (VHF analog and digital) ABC- affiliated commercial station with city of license in Manchester, NH, 1 (UHF analog and digital) Telemundoaffiliated station with city of license in Merrimack, NH, 1 (UHF analog and digital) commercial station with city of license in Derry, NH, 1 UHF analog commercial satellite of a Boston station with city of license in Concord, NH, 1 (VHF analog and digital) noncommercial station transmitting from Durham, NH, and 1 (UHF analog and digital) noncommercial station transmitting from Keene, NH

Table 1. (Continued)

State	# TV households	Counties in DMAs for which primary city is outside the state DMA: County	# TV households in DMAs for which primary city is outside the state	% of TV households in DMAs for which primary city is outside the state	Currently operating full power broadcast TV stations in DMAs for which primary city is outside the state
New Jersey	3,149,060	New York City, NY DMA: Sussex, Passaic, Bergen, Warren, Morris, Essex, Hunterdon, Somerset, Union, Middlesex, Monmouth, Ocean, Hudson	53,440+162,910+ 337,120+40,540+ 177,670+283,930+ 45,890+114,650+ 187,390+274,310+ 233,080+210,890+ 235,700+	100.00%	New York City, NY DMA: 1 UHF analog Telefutura-affiliated station with city of license in Newark, NJ, 1 (UHF analog and digital) Univision-affiliated station with city of license in Paterson, NJ, 1 UHF analog Telemundo-affiliated station with city of license in Linden, NJ, 1 VHF analog commercial station with city of license in Secaucus, NJ, 1 (UHF analog and digital) commercial station with city of license in Newton, NJ, 1 (UHF analog and digital) noncommercial station transmitting from Montclair, NJ, 1

State	# TV households	Counties in DMAs for which primary city is outside the state DMA: County	# TV households in DMAs for which primary city is outside the state	% of TV households in DMAs for which primary city is outside the state	Currently operating full power broadcast TV stations in DMAs for which primary city is outside the state
					(UHF analog and digital) noncommercial station transmitting from West Milford, NJ, 1 UHF analog non-commercial station transmitting from New Brunswick, and 1 digital noncommercial station transmitting from New York
		Philadelphia, PA DMA: Burlington, Camden, Gloucester, Salem, Cumberland, Atlantic, Cape May, Mercer	162,600+188,460+95,850+24,570+49,800+97,320+43,090+129,850= 3,149,060		Philadelphia, PA DMA: 1 (UHF analog and digital) Telemundo-affiliated station with city of license in Atlantic City, NJ, 1 UHF digital commercial station with city of license in Atlantic City, NJ, 1 UHF analog Univisionaffiliated station with city of license in Vineland, NJ, 1 UHF analog NBC-affiliated

Table 1. (Continued)

State	# TV households	Counties in DMAs for which primary city is outside the state DMA: County	# TV households in DMAs for which primary city is outside the state	% of TV households in DMAs for which primary city is outside the state	Currently operating full power broadcast TV stations in DMAs for which primary city is outside the state
New Mexico	685,270	Amarillo, TX DMA: Union, Quay, Curry, Roosevelt	1,670+4,130+16,800+ 6,470+	13.41%	Amarillo, TC DMA: 1 (VHF analog and digital) ABC-affiliated commercial satellite of an Amarillo station with city of license in Clovis, NM, and 1 (VHF analog and digital) noncommercial station transmitting from Portales, NM;
		Odessa-Midland, TX DMA: Lea South	1,770+		Odessa-Midland, TX DMA: no station with city of license in NM;
		El Paso, TX-Las Cruces, NM DMA: Dona Ana	61,050= 91,890		El Paso, TX-Las Cruces, NM DMA: 1(UHF analog and digital) Telemundo-affiliated commercial station with city of license in Las Cruces, NM, and 1 (UHF analog and digital) noncommercial station transmitting from Las Cruces, NM

State	# TV households	Counties in DMAs for which primary city is outside the state DMA: County	# TV households in DMAs for which primary city is outside the state	% of TV households in DMAs for which primary city is outside the state	Currently operating full power broadcast TV stations in DMAs for which primary city is outside the state
New York	7,025,170	Burlington, VT-Plattsburgh, NY DMA: Essex, Franklin, Clinton	15,310+17.980+ 29,880= 63,170	0.90%	Burlington, VT-Plattsburgh, NY DMA: 1 (VHF analog and digital) NBC- affiliated commercial station with city of license in Plattsburgh, NY, and 1 (UHF analog and digital) noncommercial station transmitting from Plattsburgh, NY
North Carolina	3,285,010	Atlanta, GA DMA: Clay	4,090+	12.71%	Atlanta, GA DMA: no station with city of license in NC;
		Norfolk-Portsmouth-Newport News, VA DMA: Northampton, Hertford, Gates, Camden, Pasquotank, Currituck, Perquimans, Chowan, Dare	8,900+9,140+3,930+ 2,870+13,250+7,480+ 4,810+5,590+13,570+		Norfolk-Portsmouth-Newport News, VA DMA: 1 VHF analog commercial station with city of license in Manteo, NC;

Table 1. (Continued)

State	# TV households	Counties in DMAs for which primary city is outside the state DMA: County	# TV households in DMAs for which primary city is outside the state	% of TV households in DMAs for which primary city is outside the state	Currently operating full power broadcast TV stations in DMAs for which primary city is outside the state
		Chattanooga, TN DMA: Cherokee	10,610+		Chattanooga, TN DMA: no station with city of license in NC;
		Myrtle Beach-Florence, SC DMA: Robeson, Scotland	44,720+13,680+		Myrtle Beach-Florence, SC DMA: 1 (UHF analog and digital) noncommercial station transmitting from Lumberton, NC;
		Greenville-Spartanburg-Anderson, SCAsheville, NC DMA: Graham, Swain, Haywood, Madison, Yancey, Mitchell, McDowell, Buncombe, Macon, Jackson, Transylvania, Henderson, Polk, Rutherford	3,440+5,340+24,160+8,480+7,410+6,960+17,360+87,600+13,270+14,230+13,050+39,600+8,220+25,930=417,690		Greenville-Spartanburg-Anderson, SCAsheville, NC DMA: 1 (UHF analog and digital) commercial station with city of license in Asheville, NC, and 1 (UHF analog and digital) noncommercial station transmitting from Asheville, NC

State	# TV households	Counties in DMAs for which primary city is outside the state DMA: County	# TV households in DMAs for which primary city is outside the state	% of TV households in DMAs for which primary city is outside the state	Currently operating full power broadcast TV stations in DMAs for which primary city is outside the state
North Dakota	253,780	none	0	0.00%	
Ohio	4,506,760	Charleston-Huntington, WV DMA: Athens, Meigs, Vinton, Jackson, Gallia, Lawrence, Scioto	22,330+ 9,280+4,940+12,650+ 11,960+24,970+ 30,380+	5.24%	Charleston-Huntington, WV DMA: 1 (UHF analog and digital) commercial station with city of license in Portsmouth, OH, 1 UHF analog noncommercial station transmitting from Portsmouth, OH, and 1 (UHF analog and digital) noncommercial station transmitting from Athens, OH;
		Fort Wayne, IN DMA: Paulding, Van Wert	7,900+11,550+		Fort Wayne, IN DMA: no station with city of license in OH;
		Parkersburg, WV DMA: Washington	25,270+		Parkersburg DMA: no station with city of license in OH:

Table 1. (Continued)

State	# TV households	Counties in DMAs for which primary city is outside the state DMA: County	# TV households in DMAs for which primary city is outside the state	% of TV households in DMAs for which primary city is outside the state	Currently operating full power broadcast TV stations in DMAs for which primary city is outside the state
		Wheeling, WV-Steubenville, OH DMA: Noble, Monroe, Belmont, Jefferson, Harrison	4,590+6,060+27,870+ 30,010+6,440= 236,200		Wheeling, WV-Steubenville, OH DMA: 1 (VHF analog and digital) NBC-affiliated commercial station with city of license in Steubenville, OH
Oklahoma	1,358,210	Shreveport, LA DMA: McCurtain	12,990+	16.33%	Shreveport, LA DMA: no station with city of license in OK;
		Fort Smith-Fayetteville, Springdale-Rogers, AR DMA: Sequoyah, Le Flore,	14,900+17,820+		Fort Smith-Fayetteville-Springdale-Rogers, AR DMA: no station with city of license in OK;
		Amarillo, TX DMA: Cimarron, Texas, Beaver	1,250+7,090+2,060+		Amarillo, TX DMA: no station with city of license in OK;
		Joplin, MO-Pittsburg, KS DMA: Ottawa	12,880+		Joplin, MO-Pittsburg, KS DMA: no station with city of license in OK;

State	# TV households	Counties in DMAs for which primary city is outside the state DMA: County	# TV households in DMAs for which primary city is outside the state	% of TV households in DMAs for which primary city is outside the state	Currently operating full power broadcast TV stations in DMAs for which primary city is outside the state
		Wichita Falls, TX-Lawton, OK DMA: Jackson, Tillman, Cotton, Comanche, Stephens, Jefferson	10,090+ 3,450+2,680+39,360+ 17,090+2,650+		Wichita Falls, TX-Lawton, OK DMA: 1 VHF analog ABC-affiliated commercial station with city of license in Lawton, OK;
		Sherman, TX-Ada, OK DMA: Carter, Love, Marshall, Johnston, Pontotoc, Coal, Atoka, Bryan, Choctaw, Pushmataha	17,950+3,420+ 5,510+4,030+13,970+ 2,360+5,000+14,610+ 6,080+4,540= 221,780		Sherman, TX-Ada, OK DMA: 1 (VHF analog and digital) NBC-affiliated commercial station with city of license in Ada, OK
Oregon	1,353,190	Spokane, WA DMA: Wallowa	2,910+	3.05%	Spokane, WA DMA: no station with city of license in OR;
		Boise, ID DMA: Malheur, Grant	10,350+3,040+		Boise, ID DMA: no station with city of license in OR:
		Yakima-Pasco-Richland-Kennewick, WA DMA: Umatilla	25,040= 41,340		Yakima-Pasco-Richland-Kennewick, WA DMA: 1 VHF analog FOX-affiliated commercial station with city of license in Pendleton, OR

Table 1. (Continued)

State	# TV households	Counties in DMAs for which primary city is outside the state DMA: County	# TV households in DMAs for which primary city is outside the state	% of TV households in DMAs for which primary city is outside the state	Currently operating full power broadcast TV stations in DMAs for which primary city is outside the state
Pennsylvania	4,801,400	New York City, NY DMA: Pike	18,980+	3.36%	New York City, NY DMA: no station with city of license in PA;
		Washington, DC DMA: Franklin, Fulton,	50,260+5,510+		Washington, DC DMA: no station with city of license in PA;
		Buffalo, NY DMA: McKean, Potter,	17,660+6,890+		Buffalo, NY DMA: no station with city of license in PA;
		Youngstown, OH DMA: Mercer	46,010+		Youngstown, OH DMA: no station with city of license in PA;
		Elmira, NY DMA: Tioga	15,900= 161,210		Elmira, NY DMA: no station with city of license in PA
Rhode Island	423,690	None	0	0.00%	none

State	# TV households	Counties in DMAs for which primary city is outside the state DMA: County	# TV households in DMAs for which primary city is outside the state	% of TV households in DMAs for which primary city is outside the state	Currently operating full power broadcast TV stations in DMAs for which primary city is outside the state
South Carolina	1,604,820	Charlotte, NC DMA: York, Chester, Lancaster, Chesterfield	65,870+13,070+24,050+17,380+	17.19%	Charlotte, NC DMA: 1 (UHF analog and digital) commercial station with city of license in Rock Hill, SC, and 1 UHF analog on-commercial station transmitting from Rock Hill, SC;
		Savannah, GA DMA: Hampton, Jasper, Beaufort	7,640+7,320+50,010+		Savannah, GA DMA: 1 UHF analog FOX-affiliated commercial station with city of license in Hardeeville, SC, and 1 UHF analog noncommercial station transmitting from Beuafort, SC;
		Augusta, GA DMA: McCormick, Edgefield, Aiken, Barnwell, Bamberg, Allendale	3,720+9,070+58,530+9,320+6,120+3,830= 275,930		Augusta, GA DMA: 1 UHF analog noncommercial station transmitting from Allendale, SC
South Dakota	293,510	Sioux City, IA DMA: Union	5,040+	2.15%	Sioux City, IA DMA: no station with city of license in SD;

Table 1. (Continued)

State	# TV households	Counties in DMAs for which primary city is outside the state DMA: County	# TV households in DMAs for which primary city is outside the state	% of TV households in DMAs for which primary city is outside the state	Currently operating full power broadcast TV stations in DMAs for which primary city is outside the state
		Minot-Bismarck-Dickinson, ND DMA: Corson	1,260= 6,300		Minot-Bismarck-Dickinson, ND DMA: no station with city of license in SD
Tennessee	2,297,620	Paducah, KY-Cape Girardeau-Harrisburg, MO-Mount Vernon, IL DMA: Lake, Obion, Weakley	2,260+13,330+13,810+	1.84%	Paducah, KY-Cape Girardeau-Harrisburg, MOMount Vernon, IL DMA: no station with city of license in TN;
		Huntsville-Decatur, AL DMA: Lincoln	12,820= 42,220		Huntsville-Decatur, AL DMA: no station with city of license in TN;
Texas	7,807,130	Shreveport, LA DMA: Bowie, Morris, Marion, Harrison, Panola, Shelby, Cass, Titus	34,600+5,190+4,730+ 24,370+8,810+9,680+ 12,160+9,670 = 109,210	1.40%	Shreveport, LA DMA: no station with city of license in TX
Utah	731,730	none	0	0.00%	none

State	# TV households	Counties in DMAs for which primary city is outside the state DMA: County	# TV households in DMAs for which primary city is outside the state	% of TV households in DMAs for which primary city is outside the state	Currently operating full power broadcast TV stations in DMAs for which primary city is outside the state
Vermont	243,040	Boston, MA-Manchester, NH DMA: Windham	17,770+	13.42%	Boston, MA-Manchester, NH DMA: no station with city of license in VT;
		Albany-Schenectady-Troy, NY DMA: Bennington	14,850= 32,6200		Albany-Schenectady-Troy, NY DMA: no station with city of license in VT
Virginia	2,824,170	Washington, DC DMA: Arlington, Fairfax, Loudoun, Clarke, Frederick Warren, Shenandoah, Page, Rappahannock, Culpeper, Fauquier, Prince William, Stafford, Spottsylvania, King George, Westmoreland	151,730+381,000+ 75,480+5,150+34,670+ 12,650+14,860+9,440+ 2,660+12,870+21,500+ 122,650+34,040+44,100+ 6 ,460+6,870+	38.26%	Washington, DC DMA: 1 (UHF analog and digital) commercial station with city of license in Manassas, VA, 1 UHF analog Telefuturaaffiliated commercial station with city of license in Arlington, VA, 1 (UHF analog and digital) noncommercial station transmitting from Front Royal, VA, 1 UHF analog noncommercial station transmitting from Fairfax, VA, and 1 UHF analog noncommercial station transmitting from Goldvein, VA

Table 1. (Continued)

State	# TV households	Counties in DMAs for which primary city is outside the state DMA : County	# TV households in DMAs for which primary city is outside the state	% of TV households in DMAs for which primary city is outside the state	Currently operating full power broadcast TV stations in DMAs for which primary city is outside the state
		Raleigh-Durham, NC DMA: Mecklenburg,	13,600+		Raleigh-Durham, NC DMA: no station with city of license in VA;
		Bluefield, Beckley-Oak Hill, WV DMA: Tazewell	17,870+		Bluefield-Beckley-Oak Hill, WV DMA: no station with city of license in VA
		Tri-Cities (Kingsport-Johnson City, TNBristol, VA) DMA: Buchanon, Dickinson, Lee, Russell, Scott, Smyth, Washington, Wise	9,980+6,700+9,630+ 11,710+9,950+13,670+		Tri-Cities (Kingsport-Johnson City, TNBristol, VA) DMA: 1 VHF analog and digital) NBC-affiliated commercial station with city of license in Bristol, VA, 1 UHF analog
Washington	2,306,020	Portland, OR DMA: Wahkiakum, Cowlitz, Clark, Skamania, Klickitat	1,500+35,940+134,260+3 , 470+7,340= 182,510	7.92%	Portland, OR DMA: 1 (UHF analog and digital) commercial station with city of license in Vancouver, WA

State	# TV households	Counties in DMAs for which primary city is outside the state DMA: County	# TV households in DMAs for which primary city is outside the state	% of TV households in DMAs for which primary city is outside the state	Currently operating full power broadcast TV stations in DMAs for which primary city is outside the state
West Virginia	733,120	Washington, DC DMA: Jefferson, Berkeley, Morgan, Hampshire, Mineral, Grant, Hardy	17,170+31,880+ 6,390+7,610+10,450+ 4,370+5,260+	18.00%	Washington, DC DMA: 1 (UHF analog and digital) commercial station with city of license in Martinsburg, WV;
		Pittsburgh, PA DMA: Monongalia, Preston	34,290+11,390+		Pittsburgh, PA DMA: 1 (UHF analog and digital) noncommercial station transmitting from Morgantown, WV;
		Harrisonburg, VA DMA: Pendleton	3,150= 131,960		Harrisonburg, VA DMA: no station with city of license in WV
Wisconsin	2,148,430	Minneapolis-St. Paul, MN DMA: Burnett, Washburn, Polk, Barron, St. Croix, Dunn, Pierce	6,800+6,870+17,230+ 18,160+25,950+15,020+13 ,660+	6.82%	Minneapolis-St. Paul, MN DMA: 1 UHF analog noncommercial station transmitting from Menomonie, WI;
		Marquette, MI DMA: Florence	2,170+		Marquette, MI DMA: no station with city of license in WI;

Table 1. (Continued)

State	# TV households	Counties in DMAs for which primary city is outside the state DMA: County	# TV households in DMAs for which primary city is outside the state	% of TV households in DMAs for which primary city is outside the state	Currently operating full power broadcast TV stations in DMAs for which primary city is outside the state
		Duluth, MN-Superior, WI DMA: Sawyer, Douglas, Bayfield, Ashland, Iron	6,700+18,020+6,180+ 6,570+3,170= 146,500		Duluth, MN-Superior, WI DMA: 1 (VHF analog and digital) NBC-affiliated commercial station with city of license in Superior, WI
Wyoming	195,370	Denver, CO DMA: Carbon, Albany, Platte, Niobrara, Johnson, Campbell	5,900+12,470+3,660+ 940+3,150+13,340+	54.55%	Denver, CO DMA: 1 VHF analog ABC-affiliated commercial station (a satellite of a Casper, WY station) with city of license in Rawlins, WY;
		Salt Lake City, UT DMA: Sublette, Lincoln, Uinta, Sweetwater	2,420+5,550+ 6,870+13,360+		Salt Lake City, UT DMA: 1 VHF analog CBS-affiliated station (a satellite of a Casper, WY station) with city of license in Rock Springs, WY;

State	# TV households	Counties in DMAs for which primary city is outside the state DMA: County	# TV households in DMAs for which primary city is outside the state	% of TV households in DMAs for which primary city is outside the state	Currently operating full power broadcast TV stations in DMAs for which primary city is outside the state
		Idaho Falls-Pocatello, ID DMA: Teton	8,150+		Idaho Falls-Pocatello, ID DMA: 1 VHF analog NBC-affiliated commercial station (a satellite of a Pocatello, ID station) with city of license in Jackson, WY, and 1 VHF analog commercial station with city of license in Jackson, WY;
		Billings, MT DMA: Park, Big Horn	10,420+4,130+		Billings, MT DMA: no station with city of license in WY;
		Rapid City, SD DMA: Sheridan, Crook, Weston	11,310+2,280+2,620= 106,570		Rapid City, SD DMA: 1 (VHF analog and digital) ABC-affiliated commercial satellite of a Rapid City, SD station with city of license in Sheridan, WY, and 1 VHF analog commercial station with city of license in Sheridan WY

Sources: Nielsen Media Research, *U.S. Television Household Estimates*, September 2003, for data on the number of television households in each county and the DMAs to which each county is assigned; Warren Communications News, *Television & Cable Factbook 2004*, for data on the city of license and DMA of each commercial broadcast television station and the transmitting location of each noncommercial broadcast television station.

REFERENCES

[1] 47 U.S.C. § 534. Each cable system also is required to carry the signals of certain qualified local low power television stations (47 U.S.C. § 534) and certain qualified local noncommercial television stations (47 U.S.C. § 535). Low power television (LPTV) service was created in 1982 to provide opportunities for locally-oriented television service in small communities. These communities may be in rural areas or may be individual communities within larger urban areas. LPTV stations are not considered "full-service" stations and have "secondary spectrum priority" to full-service stations. This means LPTV stations must not cause interference to the reception of existing or future full-service television stations, must accept interference from full-service stations, and must yield to new full-service stations, where interference occurs. LPTV stations are limited to an effective radiated power of 3 kilowatts for stations operating in the very high frequency (VHF) spectrum band and 150 kilowatts for stations operating in the ultra high frequency (UHF) spectrum band. See footnote 11 for a brief explanation of the differences between the VHF and UHF spectrum bands.

[2] Nielsen Media Research identifies television stations whose broadcast signals reach a specific area and attract the most viewers. According to Nielsen, "a DMA consists of all counties whose largest viewing share is given to stations of that same market area. Non-overlapping DMAs cover the entire continental United States, Hawaii, and parts of Alaska. There are currently 210 DMAs throughout the U.S." [http://www.nielsenmedia.com/ FAQ/dma_satellite%20service.htm], viewed on September 20, 2006. A very small number of counties are divided between two DMAs, typically because topographical features,

such as mountains, split the viewing patterns within the county. In addition, there are several very sparsely populated portions of Alaska that are not part of any county and not included in any DMA. Each year Nielsen reassigns a small number of counties to different DMAs, based on shifts in viewing patterns. For example, in 2003 Nielsen reassigned 24 counties to a different DMA.

[3] SHVERA passed as Title IX of the FY2005 Consolidated Appropriations Act (H.R. 4818, P.L. 108-447).

[4] These statutory restrictions appear in the Satellite Home Viewer Improvement Act, which is Title I of the Intellectual Property and Communications Omnibus Reform Act of 1999, included by cross reference in the FY2000 Consolidated Appropriations Act, P.L. 106-113.

[5] Under another provision of SHVIA, subscribers who are not able to receive an over the air broadcast signal of acceptable quality using a conventional, stationary rooftop antenna are eligible to receive distant television signals from their satellite provider. Under certain circumstances, these distant signals may be from stations located in the same state as the subscriber. This situation is discussed in greater detail later in this report.

[6] The definition of "significantly viewed" signals is discussed in greater detail below in the section on Cable Television. The statutory instructions on how to implement these provisions of SHVERA are discussed in the section on Satellite Television.

[7] These state-specific exceptions are discussed in greater detail below in the section on Satellite Television.

[8] Nielsen Media Research, *U.S. Television Household Estimates*, September 2003, which presents data on the number of television households in each county and the DMA to which each county is assigned.

[9] Warren Communications News, *Television & Cable Factbook 2004*, which presents data on the city of license and DMA of each commercial broadcast television station and the transmitting location of each noncommercial broadcast television station.

[10] The *Television & Cable Factbook 2004* was published in April 2004 and thus does not include stations that have begun operation in 2004. As explained later in this report, the industry currently is in the midst of a congressionally-mandated transition from the analog transmission of broadcast signals to digital transmission. While most licensees had already begun transmitting digital as well as analog signals prior to

2004, many licensees began dual transmission in 2004 and those new digital transmissions are not reflected in the data (and hence not reflected in Table 1). Since most digital transmissions have simply duplicated analog transmissions, however, this should not result in an underestimate of the amount of programming available to television households.

[11] Analog broadcast television service is provided over two portions of the radio spectrum — the very high frequency (VHF) portion and the ultra high frequency (UHF) portion. The transmission characteristics of the spectrum is such that VHF signals transmit further and require less power and therefore VHF stations tend to have a larger reach, with better reception quality and lower costs. These differences disappear when the signals are received via cable or satellite service rather than over the air.

[12] The other two are diversity of voices and competition.

[13] The source for this discussion of broadcasters' public interest obligations is the introduction to the FCC's Notice of Inquiry, *In the Matter of Broadcast Localism*, MB Docket No. 04-233, adopted June 7, 2004 and released July 1, 2004, 1-5. To date, the FCC has not adopted any rules or taken any action in this proceeding.

[14] 47 U.S.C. § 309(a). This concept of public trusteeship was reiterated by the Commission in *Advanced Television Systems and Their Impact upon the Existing Television Broadcast Service*, 12 FCC Rcd 12829 (1977), in which it noted that even as they transition to digital technology, "broadcasters will remain trustees of the public's airwaves."

[15] 47 U.S.C. § 307(b).

[16] See Amendment of Section 3.606 of the Commission's Rules and Regulations, 41 F.C.C. 148, 167 (1952). The Commission's first television allocation priority is to "provide at least one television station to all parts of the United States"; its second is to "provide each community with at least one television broadcast station."

[17] Pacific Broadcasting of Missouri LLC, 18 FCC Rcd 2291 (2003) (quoting Public Service Broadcasting of West Jordan, Inc., 97 F.C.C. 2d 960, 962 (Rev. Bd. 1984)).

[18] FCC v. Allentown Broadcasting Corp., 349 U.S. 358, 362 (1955).

[19] 47 C.F.R. § 73.685(a).

[20] 47 C.F.R. § 73.1125.

[21] Amendment of Sections 73.1125 and 73.1130 of the Commission's Rules, the Main Studio and Program Origination Rules for Radio and Television Broadcast Stations, 3 FCC Rcd 5024, 5026 24 (1988).

[22] 47 C.F.R. § 73.352(e)(11)(i). These lists must be retained until final action has been taken on the station's renewal application.

[23] The Grade A contour around a station's transmitter identifies the geographic area in which satisfactory service is expected at least 90% of the time for at least 70% of the receiving locations. The Grade B contour identifies the geographic area in which the quality of picture is expected to be satisfactory to the median observer at least 90% of the time for at least 50% of the receiving locations within the contours, in the absence of interfering co-channel and adjacent-channel signals. (See Warren Communications News, *Televison & Cable Factbook 2004*, volume 72, at p. A-14.)

[24] See Notice of Inquiry, *In the Matter of Broadcast Localism*, MB Docket No. 04-233, adopted June 7, 2004 and released July 1, 2004, at 3 and footnote 11.

[25] For example, in Re Application of WHYY, Inc., for Renewal of License for Noncommercial Educational Television Station WHYY-TV, Wilmington Delaware, the Commission stated "Although the petitioners emphasize the station's primary obligation to serve the needs of Wilmington, its city of license, WHYY-TV owes a secondary duty to serve other nearby areas, which include Philadelphia and Camden and Trenton, New Jersey, as we have previously recognized. 18 RR 2d 1603 (1970)." 53 F.C.C. 2d 421 (para. 9). In a subsequent decision involving the same station, the Commission expanded on this: "Although WHYY believes that it is a television station licensed to serve Wilmington and '[a]lmost equally important ... adjacent metropolitan areas of Philadelphia and Camden,' the licensee's first and primary obligation is to serve the local needs and interests of community of license — Wilmington. This primary obligation to Wilmington, contrary to WHYY's assertions, has been emphasized by the Commission for at least the last twenty-three years.... While regional programs can address the interests of Wilmington residents, such programs cannot serve other service area residents to the detriment of the citizens of Wilmington. The licensee's prime and most important focus must be on the problems, needs, and interests of its community of license. However, as we outlined in our 1975 decision to renew the license of WHYY, 53 FCC 2d 421 (1975), while the station's primary obligation is to serve the needs of its city of

license, WHYY also has a secondary duty to serve other nearby areas including Philadelphia, Camden, and Trenton, New Jersey." 93 F.C.C. 2d 1096 (para. 20) (1983), emphasis in original.

[26] In an order reallocating channel 9 from New York City to Secaucus, New Jersey, (Channel 9 Reallocation (WOR-TV), 53 RR 2d 469 (1983)), the Commission stated: "It is expected that the licensee will devote itself to meeting the special needs of its new community (and the needs of the northern New Jersey area in general).... In the usual case, Secaucus, the city of assignment, would be the primary focus of the licensee's programming responsibilities. However, we have previously determined that the lack of local VHF television service to this highly populated area of northern New Jersey presented a unique set of circumstances (See, e.g, Docket No. 20350, 2nd R&O, 59 FCC 2d 1386 [37 RR 2d 1275] (1976), wherein special service obligations have been imposed on all New York City and Philadelphia TV stations. Accordingly, we expect RKO to perform a higher degree of service to its Grade B coverage area than is normally required of a broadcast licensee. At renewal time RKO will be judged by how it met the obligation to serve the greater service needs of northern New Jersey, which we view as broader than the specific needs of Secaucus."

[27] The data underlying the following discussion are found in Table 1.

[28] Interestingly, four stations were listed under New York State in the *Television & Cable Factbook 2004*, a data source widely used in the industry, despite having city of license in New Jersey. In the 2005 edition, three of those stations were listed under New York State. In the 2006 edition, all four of the stations were listed under New Jersey.

[29] Today, upwards of 85% of all U.S. television households receive their broadcast signals by a means other than over the air reception. According to data presented by the National Cable and Telecommunications Association on its website ([http://www.ncta.com], Statistics and Resources, viewed on September 20, 2006), in June 2006 there were 110.9 million U.S. television households, of which 65.5 million subscribed to cable television and 28.0 million subscribed to satellite television or some other non-cable multichannel video program service. (Adding these two figures together would create a slight amount of double counting of non-broadcast households as a small portion of these households subscribed to both cable and a non-cable service). The statutory, regulatory, and private contractual restrictions on cable and satellite

systems carrying the signals of broadcast stations located in-state, but outside-the-DMA are discussed in the cable and satellite sections of this report.

[30] *Cable Television Report and Order*, adopted on 2/2/72, 36 FCC 2nd 143 (1972).

[31] MVPDs provide packages of video programming to subscribers for a monthly fee. The overwhelming majority of television households that receive their programming from MVPDs subscribe to cable or direct broadcast satellite systems, but a small number of households get low power "C-band" home satellite dish (HSD) service, wireless cable service such as multichannel multipoint distribution service (MMDS), or service provided by municipal or private overbuilding broadband service providers (BSP) or by private cable operators. See Federal Communications Commission, *Annual Assessment of the Status of Competition in the Market for the Delivery of Video Programming, Twelfth Annual Report*, adopted February 10, 2006, released March 3, 2006.

[32] Codified at 47 U.S.C. §§ 534 and 535.

[33] As explained above, each broadcast television station can choose, once every three years, between two options: (1) negotiating a retransmission consent agreement with each local cable operator to make its programming available in exchange for compensation or (2) requiring the local cable operator to carry its programming at no charge to the cable operator.

[34] A noncommercial educational station that places a Grade B signal over a cable system's principal headend, or whose city of license is within fifty miles of the cable system's principal headend, is considered "local" for this purpose.

[35] FCC Fact Sheet, "Cable Television," section entitled "Signal Carriage Requirements," dated June 2000, available at [http://www.fcc.gov/mb/facts/csgen.html], viewed on 9/20/2006. Television markets originally were defined according to Arbitron market definitions, but when Arbitron discontinued performing this service, the FCC chose to use the Nielsen DMAs.

[36] Commercial television station licensees are entitled to protect the network programming they have contracted for by exercising non-duplication rights against more distant television broadcast stations carried on a local cable television system that serves more than 1,000 subscribers. Commercial broadcast stations may assert these non-duplication rights regardless of whether or not their signals are being

transmitted by the local cable system and regardless of when, or if, the network programming is scheduled to be broadcast. Generally, the zone of protection for such programming cannot exceed 35 miles for stations licensed to a community in the Commission's list of top 100 television markets or 55 miles for stations licensed to communities in smaller television markets. In addition, a cable operator does not have to delete the network programming of any station which the Commission has previously recognized as significantly viewed in the cable community.

[37] With respect to non-network programming, cable systems that serve at least 1,000 subscribers may be required, upon proper notification, to provide syndicated protection to broadcasters who have contracted with program suppliers for exclusive exhibition rights to certain programs within specific geographic areas, whether or not the cable system affected is carrying the station requesting this protection. However, no cable system is required to delete a program broadcast by a station that either is significantly viewed or places a Grade B or better contour over the community of the cable system.

[38] A cable system located within 35 miles of the city of license of a broadcast station where a sporting event is taking place may not carry the live television broadcast of the sporting event on its system if the event is not available live on a local television broadcast station, if the holder of the broadcast rights to the event, or its agent, requests such a blackout. The holder of the rights is responsible for notifying the cable operator of its request for program deletion at least the Monday preceding the calendar week during which the deletion is desired. If no television broadcast station is licensed to the community in which the sports event is taking place, the 35-mile blackout zone extends from the broadcast station's licensed community with which the sports event or team is identified. If the event or local team is not identified with any particular community (for instance, the New England Patriots), the 35-mile blackout zone extends from the community nearest the sports event which has a licensed broadcast station. The sports blackout rule does not apply to cable television systems serving less than 1,000 subscribers, nor does it require deletion of a sports event on a broadcast station's signal that was carried by a cable system prior to March 31, 1972. The rule does not apply to sports programming carried on non-broadcast program distribution services such as ESPN. These services, however, may be subject to private contractual blackout restrictions.

[39] 17 U.S.C. § 111.
[40] Codified at 47 U.S.C. § 534.
[41] 16 FCC Rcd 16099 (2001).
[42] For a full discussion of this transition, see CRS Report RL31260, *Digital Television: AnOverview*, by Lennard Kruger.
[43] *In the Matter of Carriage of Digital Television Broadcast Signals: Amendments to Part 76 of the Commission's Rules*, CS Docket No. 98-120, Second Further Notice of Proposed Rulemaking, adopted April 25, 2007 and released May 4, 2007.
[44] 47 U.S.C. § 534(b)(7).
[45] See the Statement of Commissioner Jonathan S. Adelstein and the Statement of Commissioner Robert M. McDowell, *re: Cable Carriage of Digital Television BroadcastSignals* (CS Docket No. 98-120), April 25, 2007.
[46] *In the Matter of Carriage of Digital Television Broadcast Signals: Amendments to Part 76 of the Commission's Rules,* CS Docket No. 98-120, Second Report and Order and First Order on Reconsideration, adopted February 10, 2005, released February 23, 2005, at 33.
[47] P.L. 98-549, 47 U.S.C. 531 (Section 611 of the Communications Act).
[48] SHVIA is Title I of the Intellectual Property and Communications Omnibus Reform Act of 1999, included by cross reference in the FY2000 Consolidated Appropriations Act, P.L. 106-113. For more information on SHVIA and related issues, see CRS Report RS21768, Satellite Television: Reauthorization of the Satellite Home Viewer Improvement Act (SHVIA) — Background and Key Issues, by Marcia S. Smith, and CRS Report RS20425, Satellite Television: Historical Information on SHVIA and LOCAL, by Marcia S. Smith.
[49] 17 U.S.C. 122.
[50] FCC Information Sheet, "Television Broadcast Channels on Satellite," dated May 2006, available at [http://www.fcc.gov] (under Media Bureau and Information Sheet on Broadcast Signals on DBS), viewed on October 12, 2006.
[51] The "distant network signal" license originated in the 1988 Satellite Home Viewer Act and was extended in 1994 and in the 1999 SHVIA. See 17 U.S.C. 119.
[52] P.L. 108-447, Title IX, Satellite Home Viewer Extension and Reauthorization Act of 2004, Title II, Federal Communications Commission Operations, Sec. 202 (47 U.S.C. 340(a)). 47 U.S.C. 340(b) lays out, separately for analog service and for digital service, certain limitations on the signals that may be carried. It also excludes

those limitations when a subscriber is located in a local market (i.e., DMA) in which there are no network stations affiliated with the same television network as the station whose signal is being retransmitted pursuant to the section, and provides for a process by which a satellite company can seek a waiver of the limitations.

[53] In the Matter of Implementation of the Satellite Home Viewer Extension and Reauthorization Act of 2004 and Implementation of Section 340 of the Communications Act, MB Docket No. 05-49, Report and Order, adopted November 2, 2005, released November 3, 2005, Appendix C.

[54] P.L. 108-447, Title IX, Satellite Home Viewer Extension and Reauthorization Act of 2004, Title I, Statutory License for Satellite Carriers, Sec. 102 (17 U.S.C. 119(a)(3)). Sec. 102 also includes a limitation and a process for waiving the limitation. This royalty-free compulsory copyright license is not permanent; it will have to be renewed at the end of 2009.

[55] *In the Matter of Implementation of the Satellite Home Viewer Extension and Reauthorization Act of 2004, Procedural Rules*, Order, adopted March 28, 2005, released March 30, 2005, implementing P.L. 108-447, Title IX, Satellite Home Viewer Extension and Reauthorization Act of 2004, Title II, Federal Communications Commission Operations, Sec. 202 (47 U.S.C. 340(h)(1)).

[56] P.L. 108-447, Title IX, Satellite Home Viewer Extension and Reauthorization Act of 2004, Title I, Statutory License for Satellite Carriers, Sec. 102 (17 U.S.C. 119(a)(2)(C)(i)).

[57] P.L. 108-447, Title IX, Satellite Home Viewer Extension and Reauthorization Act of 2004, Title I, Statutory License for Satellite Carriers, Sec. 102 (17 U.S.C. 119(a)(2)(C)(ii)).

[58] P.L. 108-447, Title IX, Satellite Home Viewer Extension and Reauthorization Act of 2004, Title I, Statutory License for Satellite Carriers, Sec. 102 (17 U.S.C. 119(a)(2)(C)(iii)).

[59] P.L. 108-447, Title IX, Satellite Home Viewer Extension and Reauthorization Act of 2004, Title II, Federal Communications Commission Operations, Sec. 211, Carriage of Television Signals to Certain Subscribers (47 U.S.C. 341). If the cable operator or satellite carrier is authorized to carry less than three broadcast station signals, then it may elect to retransmit up to two broadcast signals. As a consequence of this provision, cable and satellite carriers can: provide subscribers in Grant County the signals of KGW (NBC) and KOPB (PBS) from Portland, Oregon; provide subscribers in Malheur County

the signals of KGW (NBC) from Portland, Oregon and KTVR (PBS) from La Grande, Oregon; provide subscribers in Umatilla County the signals of KATU (ABC), KGW (NBC), KOIN (CBS), KPTV (FOX), and KOPB (PBS) from Portland, Oregon, KFFX (FOX) of Pendleton, Oregon, and KTVR (PBS) from La Grande, Oregon; and provide subscribers in Wallowa County the signals of KGW (NBC) and KPTV (FOX) from Portland, Oregon, and KTVR (PBS) from La Grande, Oregon.

[60] P.L. 108-447, Title IX, Satellite Home Viewer Extension and Reauthorization Act of 2004, Title I, Statutory License for Satellite Carriers, Sec. 102 (17 U.S.C. 119(a)(2)(iv).

[61] P.L. 108-447, Title IX, Satellite Home Viewer Extension and Reauthorization Act of 2004, Title I, Statutory License for Satellite Carriers, Sec. 111(b) (17 U.S.C. 119(a)(16).

[62] P.L. 108-447, Title IX, Satellite Home Viewer Extension and Reauthorization Act of 2004, Title II, Federal Communications Commission Operations, Sec. 210, Satellite Carriage of Television Stations in Noncontiguous States (47 U.S.C. 338(a)(4)). Within one year of passage of SHVERA, the broadcast stations must choose between making their signals available under the terms of retransmission consent or under mandatory carriage.

[63] These listings were as follows:

Station	City of License	State Listed in 2004	City Listed in 2004	State Listed in 2005	City Listed in 2005	State Listed in 2006	City Listed in 2006
WFUT	Newark, NJ	New York	New York-Newark, NJ	New York	New York-Newark, NJ	New Jersey	Newark
WNJU	Linden, NJ	New York	New York-Newark, NJ	New Jersey	New York-Newark, NJ	New Jersey	Linden
WXTV	Paterson, NJ	New York	New York-Paterson, NJ	New York	New York-Paterson, NJ	New Jersey	Paterson
WWOR	Secaucus, NJ	New York	New York-Secaucus, NJ	New York	New York-Secaucus, NJ	New Jersey	Secaucus
WCTV	Thomasville, GA	Florida	Tallahassee, FL-	Florida	Tallahassee, FL-	Georgia	Thomasville

			Thomas- ville, GA		Thomas- ville, GA		
WRBU	East St. Louis, IL	Missouri	St. Louis-East St. Louis, IL	Missouri	St. Louis-East St. Louis, IL	Illinois	East St. Louis
KBJR	Superior, WI	Minnesota	Duluth-Superior, WI	Wisconsin	Duluth-Superior, WI	Wisconsin	Superior

[64] Notice of Inquiry, *In the Matter of Broadcast Localism*, MB Docket No. 04-223, adopted June 7, 2004 and released July 1, 2004. Initial comments are due on November 1, 2004.

[65] Commission decisions to date do not provide explicit guidance. For example, in the FCC license renewal decision most on point here, *In re Application of WHYY, Inc. for Renewal of License of Station WHYY-TV, Wilmington, Delaware*, 93 F.C.C. 2d 1086 (1983), the Broadcast and Communications Commission of the City Council of Wilmington, Delaware challenged renewal of the license, alleging that the station broadcast more programming that focused on Philadelphia than on Wilmington. The Commission found that over its "last license term, WHYY provided an average of less than 3-3 ½ hours per week of programming exclusively addressed to the needs and interests of Delaware residents. Thus, it would appear there has been an erosion in the commitments which led the Commission to grant the construction permit application of WHYY. This erosion, however, does not indicate that WHYY has failed to fulfill its obligations and to treat Wilmington as its primary service area. As the Commission has previously stated, 'licensees are not bound to strict, inflexible adherence to program proposals, but are afforded broad discretion in the manner in which they respond to community problems.' *Educational Broadcasting Corporation* ... 31 FCC 85 (1961)." (93 F.C.C. 2d 1095). In the decision, the Commission goes on to state: "Programming which reflects service to Wilmington as WHYY's primary service area is not limited to programming which exclusively involves Delaware persons and issues. Regional, national and international topics may be of interest to residents of Wilmington as well as programs specifically designed for Delaware. The interests of Delaware residents, it can be concluded, flow beyond the confines of the borders of Delaware to topics of interest outside the state. Such interests are addressed by such national programs as the *McNeil-Lehrer Report, Wall Street Week, Over Easy,* and *The Advocates*, which are broadcast by the licensee.

However, despite the wealth of national programs, an educational licensee must also provide a local program service which addresses the unique problems, needs and interests of the community it is licensed to serve." Id. at 1095-1096.

[66] Public Interest Obligations of TV Broadcast Licensees, 14 FCC Rcd 21633 (1999), commonly known as the "DTV Public Interest NOI."

[67] See, for example, Standardized and Enhanced Disclosure Requirements for Television Broadcast Licensee Public Interest Obligations, 15 FCC Rcd 19186 (2000); Children's Television Obligations of Digital Television Broadcasters, 15 FCC Rcd 22946 (2000); Second Periodic Review of the Commission's Rules and Policies Affecting the Conversion to Digital Television, 18 FCC Rcd 1279 (2003); FCC Press Release, "FCC Adopts Children's Programming Obligations for Digital Television Broadcasters," Report and Order FCC 04-221, MM Docket 00-167, adopted and announced September 9, 2004.

[68] Notice of Inquiry, *In the Matter of Broadcast Localism*, MB Docket No. 04-223, adopted June 7, 2004 and released July 1, 2004, at paragraph 8.

[69] But Supreme Court rulings relating to First Amendment constraints on government regulation of broadcast stations have set heightened scrutiny when the speech to be regulated is content-based rather than content-neutral (*Turner Broadcasting Sys. v. F.C.C.*, 512 U.S. 622 (1994) at 642-3).

[70] With respect to copyright, some observers claim that the carriage of in-state programming would be fostered by expanding the congressionally mandated royalty-free compulsory license to include the secondary transmission of signals of those stations located in the same state, but outside the DMA, of the cable franchise, but that are not "significantly viewed" by television households in the county in which the cable franchise is located. Copyright holders argue, however, that now that cable is no longer an infant industry it is inappropriate to maintain the current royalty-free compulsory license, no less expand its scope.

INDEX

A

ABC, 8, 9, 27, 48, 62, 71, 73, 77, 83, 95, 107
access, ix, 4, 11, 19, 36, 38
advertising, 30
affiliates, 1
agent, 104
Alabama, 41
Alaska, x, 3, 26, 29, 38, 42, 98
antenna, 98
ants, 31
area hubs, 2
Arizona, 42
Arkansas, ix, 4, 10, 24, 42
assignment, 39, 101
Athens, 80
attention, 16
authority, 37
availability, 29

B

bandwidth, 25
beams, 25
behavior, 13
blackouts, 14, 19
Boston, ix, 9, 73, 89

broadband, 103
broadcast television, ix, 2, 3, 4, 6, 7, 8, 11, 12, 20, 23, 24, 25, 26, 28, 31, 32, 34, 35, 37, 38, 95, 99, 103
broadcaster(s), ix, 1, 5, 11, 12, 15, 16, 17, 18, 20, 21, 24, 27, 31, 34, 35, 36, 39, 99, 100, 104

C

cable operators, x, 12, 17, 18, 19, 20, 24, 35, 36, 37, 39, 103
cable service, 12, 13, 102, 103
cable system, 2, 3, 11, 12, 13, 14, 15, 16, 17, 18, 19, 20, 21, 22, 36, 37, 39, 97, 103, 104
cable television, 11, 23, 102, 104, 105
California, 44
campaigns, 30
capacity, 19, 25
carbon, 93
carrier, 26, 27, 28, 29, 39, 107
CBS, 8, 45, 50, 51, 59, 67, 68, 71, 107
central city, 7
channels, 5, 13, 18, 19, 20, 23, 36, 38
Cheyenne, 10, 70, 71
Chicago, ix, 10, 54
children, 35
Cincinnati, ix, 10, 55, 59
circulation, 13

Colorado, 39, 45
commercial, 2, 9, 10, 11, 13, 15, 17, 18, 19, 27, 28, 41, 43, 45, 46, 47, 48, 49, 50, 51, 52, 53, 54, 55, 56, 57, 59, 60, 61, 62, 63, 66, 67, 68, 71, 72, 73, 74, 75, 76, 77, 78, 80, 82, 83, 84, 86, 89, 90, 91, 93, 94, 95, 99
Communications Act, 2, 5, 13, 105, 106
community(ies), 1, 2, 5, 6, 13, 16, 19, 24, 26, 31, 32, 34, 97, 100, 101, 104, 105, 109
compensation, 12, 27, 103
competition, 23, 99
compilation, 3
Congress, vii, 1, 2, 3, 11, 13, 23, 31, 34, 36, 37, 38, 39
Connecticut, ix, 8, 10, 16, 46
consent, 11, 12, 14, 27, 28, 29, 103, 108
constraints, 25, 110
construction, 109
consumers, 2, 38
contracts, 12, 15
control, 19
conversion, 35
copyright, x, 15, 20, 23, 27, 28, 29, 36, 37, 40, 106, 110
costs, 30, 99
coverage, 9, 16, 21, 31, 102
covering, 27, 31
crime, 31
CRS, x, 7, 21, 33, 105, 106
CT, 46
customers, 24, 29

D

decisions, 13, 18, 109
Deficit Reduction Act, 17
definition, 13, 36, 98
Delaware, ix, 9, 109
demand, 37
density, 42
digital television, 17, 18, 35
distribution, 5, 12, 103, 105

District of Columbia, 47
diversity, 99
double counting, 102
duplication, 14, 20, 26, 27, 28, 29, 36, 37, 40, 104

E

economic, 2, 34
election, 11, 12, 27
England, 105
equipment, 18, 19
erosion, 109
Eureka, 43, 72
evidence, 16
evolution, 12

F

failure, 34
Federal Communications Commission, (FCC), ix, 5, 7, 8, 14, 15, 17, 18, 26, 27, 28, 29, 31, 32, 34, 35, 36, 37, 39, 40, 99, 100, 101, 102, 103, 105, 106, 107, 108, 109, 110
fee(s), 15, 20, 36, 103
First Amendment, 17, 110
flexibility, 15, 36, 37
flow, 110
football, 1
Fox, 8, 45
franchise, 20, 110
fulfillment, 16

G

games, 1
Georgia, 49, 108
goals, 37
government, 2, 19, 110
government intervention, 2
groups, 19
guidance, 7, 34, 109

H

harm, 12
Hawaii, x, 3, 26, 29, 38, 51, 98
head, 19
high definition television, 36
household, 34
households, ix, 1, 2, 3, 4, 8, 9, 10, 11, 12, 13, 16, 17, 23, 25, 26, 36, 41, 42, 43, 45, 46, 48, 49, 50, 52, 53, 54, 56, 57, 58, 60, 61, 62, 64, 65, 66, 67, 69, 70, 71, 72, 74, 75, 76, 78, 79, 80, 82, 83, 85, 86, 87, 88, 90, 91, 93, 94, 95, 99, 102, 103, 111

I

Idaho, ix, 10, 51, 94
Illinois, ix, 8, 10, 52, 109
implementation, 36
incentives, 31, 32
inclusion, 21
Indiana, ix, 10, 54
industry, 11, 14, 17, 33, 99, 102, 111
in-state, x, 3, 4, 15, 20, 21, 24, 26, 30, 102, 110
intensity, 24
interaction, 6
interference, 97
international, 110
intervention, 2

J

Jefferson, 53, 55, 82, 83, 91
Jordan, 100
jurisdiction, 22

K

Kentucky, ix, 8, 10, 59
King, 89

L

Lafayette, 44, 66
language, 9, 13, 15
law, 15, 23, 24, 25, 38
laws, 11, 37
legislation, 32
license fee, 36
licenses, 5, 7, 8, 32
likelihood, 29
limitation, 106
local community, 1, 6
local government, 31
local television stations, 17, 29
local-into-local, x, 3, 23, 24, 25, 26, 29
location, 3, 4, 14, 95, 99
Louisiana, 28, 61
low power, 13, 19, 97, 103

M

Madison, 52, 67, 71, 80
Maine, 61
management, 6
mapping, 2
market, ix, 2, 12, 13, 15, 18, 22, 23, 26, 27, 32, 34, 98, 103, 106
market incentives, 32
market share, 12
markets, ix, 2, 10, 13, 14, 15, 23, 29, 33, 103, 104
Maryland, 9, 14, 22, 61
Massachusetts, 62
media, 31
median, 100
metropolitan area, 1, 8, 10, 101
Mexico, 77
Miami, 57
Minnesota, 64
minority, 13
Mississippi, x, 3, 26, 28, 38, 39, 42, 65, 67
Missouri, ix, 4, 8, 10, 67, 100, 109

Monroe, 44, 52, 59, 68, 82
Montana, 69
Moscow, 51
mountains, 98
multichannel video programming distributor, 13

N

natural, 7
NBC, 8, 61, 76, 78, 82, 83, 90, 93, 94, 107
Nebraska, 70
negotiating, 11, 103
network, x, 1, 9, 12, 14, 20, 24, 26, 27, 28, 39, 40, 104, 106
Nevada, 44, 72
New England, 105
New Jersey, ix, 8, 9, 74, 101, 102
New Orleans, 65
New York, ix, 8, 10, 46, 74, 75, 78, 85, 101, 102
news coverage, 16
Newton, 43, 54, 74
Nielsen, x, 2, 3, 14, 23, 29, 95, 97, 98, 103
nonduplication, x
Norfolk, 71, 78

O

obligation, ix, 4, 7, 8, 22, 31, 101, 102
Ohio, 10, 55, 80
Oklahoma, 39, 82
Oneida, 51
operator, 13, 14, 19, 20, 21, 24, 25, 26, 28, 37, 103, 104, 107
OR, 44, 84, 91
Oregon, x, 3, 26, 28, 38, 39, 84, 107
Ottawa, 83
out of state, x, 20

P

Pacific, 100
Pennsylvania, 85
periodic, 35
permit, 39, 109
petitioners, 101
Philadelphia, ix, 8, 9, 47, 76, 101, 102, 109
political, 22, 30
politics, 22
population, ix, 7, 9, 10, 34
power, 2, 4, 5, 6, 13, 19, 31, 41, 42, 43, 45, 46, 48, 49, 50, 52, 53, 54, 56, 57, 58, 60, 61, 62, 64, 65, 66, 67, 69, 70, 71, 72, 74, 75, 76, 78, 79, 80, 82, 83, 85, 86, 87, 88, 90, 91, 93, 94, 97, 99, 103
preference, 22
procedures, 26
production, 6, 19
program, 6, 12, 14, 18, 19, 20, 23, 102, 104, 109
programming, x, 1, 5, 9, 11, 12, 14, 18, 19, 20, 22, 24, 25, 27, 29, 34, 35, 36, 37, 38, 99, 101, 103, 104, 105, 109, 110
promote, 23
public, ix, 5, 6, 7, 17, 18, 19, 30, 35, 37, 39, 99, 100
public interest, ix, 5, 7, 35, 37, 99
public policy, 17, 30
public welfare, 39
publishers, 34

R

radio, 5, 6, 99
radius, 7
reception, 97, 99, 102
recognition, 6
reconcile, 31
regional, 31, 101

Index

regulation, 15, 110
regulations, 40
regulatory requirements, 31
returns, 17
Rhode Island, 86
risk, 34
rivers, 7
royalty, x, 15, 23, 27, 28, 29, 37, 39, 106, 110
rural areas, 97
Rutherford, 80

S

safeguard, 12
satellite, iv, x, 1, 3, 9, 12, 23, 24, 25, 26, 27, 28, 29, 30, 31, 38, 39, 45, 62, 63, 65, 67, 71, 73, 77, 93, 94, 95, 98, 99, 102, 103, 106, 107
Satellite Home Viewer Extension and Reauthorization Act, x, 3, 26, 106, 107, 108
Satellite Home Viewer Improvement Act, 23, 98, 106
satellite service, 3, 9, 99
self-interest, 32
service provider, 103
SHVERA, 3, 26, 27, 29, 30, 38, 39, 98, 108
SHVIA, 23, 24, 98, 105, 106
signals, x, 1, 2, 3, 6, 8, 11, 12, 13, 14, 15, 17, 19, 20, 21, 23, 24, 25, 26, 27, 28, 29, 30, 31, 36, 37, 38, 39, 97, 98, 99, 100, 102, 104, 106, 107, 108, 110
South Carolina, 86
spectrum, 1, 5, 17, 18, 35, 97, 99
speech, 110
sports, x, 14, 20, 22, 24, 40, 105
St. Louis, ix, 10, 52, 109
standards, 6, 19, 26
state borders, ix, 2, 7, 8, 31, 35
statutes, 2

subscribers, x, 3, 18, 21, 22, 23, 24, 26, 28, 29, 30, 37, 38, 39, 98, 103, 104, 105, 107
suburbs, 7
suppliers, 104
Supreme Court, 6, 110
Switzerland, 55
syndicated, x, 14, 20, 26, 40, 104
systems, 3, 11, 12, 13, 14, 15, 16, 17, 18, 19, 31, 36, 102, 103, 104, 105

T

technological, 25
technology, 17, 18, 36, 100
television, ix, 1, 2, 3, 4, 6, 7, 8, 9, 10, 11, 12, 13, 14, 15, 16, 17, 18, 19, 20, 22, 23, 24, 25, 26, 28, 29, 31, 32, 33, 34, 35, 36, 37, 38, 39, 95, 97, 98, 99, 100, 101, 102, 103, 104, 106, 111
television stations, ix, 2, 3, 4, 7, 8, 9, 13, 14, 15, 17, 19, 20, 24, 25, 26, 29, 31, 32, 33, 34, 36, 97
television viewing, 8
Tennessee, 4, 88
Texas, 83, 88
threshold, 12
Title III, 5
traffic, 22
training, 19
transition, 17, 18, 35, 36, 99, 100, 105
transmission, 5, 6, 15, 17, 18, 23, 27, 28, 29, 35, 37, 99, 110
transmission capability, 6
transportation, 31
Transylvania, 80

U

UHF, 4, 8, 9, 10, 41, 43, 45, 46, 47, 48, 49, 50, 51, 52, 53, 54, 55, 56, 57, 59, 60, 61, 62, 64, 66, 67, 71, 73, 74, 75,

76, 77, 78, 79, 80, 86, 87, 89, 90, 91, 92, 97, 99
United States, 2, 98, 100
urban areas, 97
Utah, 88

V

Vermont, x, 3, 26, 28, 38, 39, 89
VHF, 8, 9, 45, 50, 51, 53, 56, 59, 63, 65, 67, 68, 70, 71, 72, 73, 74, 77, 78, 82, 83, 84, 90, 93, 94, 95, 97, 99, 102
video, 13, 18, 102, 103
video programming, 13, 103
viewing patterns, 1, 2, 13, 15, 16, 22, 98

Virginia, 9, 89, 91

W

Washington, 9, 12, 14, 22, 52, 55, 61, 81, 85, 89, 90, 91
wealth, 110
welfare, 39
Wisconsin, 92

Y

yield, 97